How to design a Gravity Flow Water System

Through worked exercises

First English Edition.
September 2010.

Santiago Arnalich

How to design a Gravity Flow Water System

Through worked exercises

First English Edition.
September 2010.

ISBN: 978-84-614-3744-3

Cover photos: Rope and bucket. Illustration page 5: Arantxa Osés Alvarez

Errata at: www.arnalich.com/dwnl/xligraxen.doc

Revision: Oliver Style and Amelia Jiménez Martín.
Translation: Oliver Style.

arnalich

water and habitat

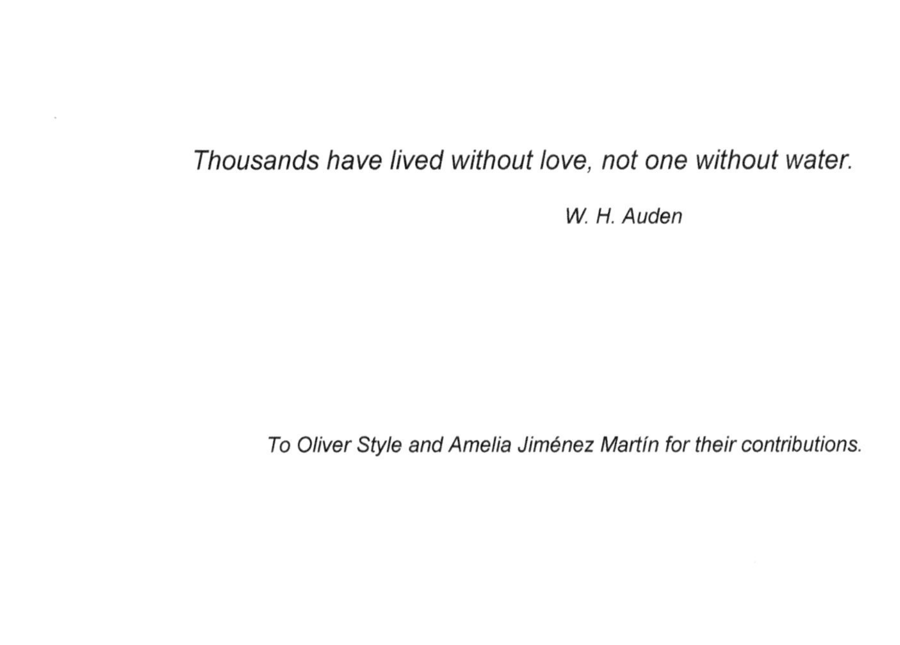

Thousands have lived without love, not one without water.

W. H. Auden

To Oliver Style and Amelia Jiménez Martín for their contributions.

Index

1. Introduction

1. 1 WHAT DO I BUILD? IS IT GOING TO WORK?

This book intends to provide you with the tools you need to answer one of the following questions:

I have a certain number of people who need water: **What do I need to build?**
I have a design proposal for a system: **Is it going to work?**

It's common in cooperation projects to underestimate gravity flow systems and think that they'll work just on their own. Saying that a gravity flow system is going to work simply because the water is flowing downhill, is like saying that a plane that's losing altitude is just going to land on its own: you'd better not be onboard! In both cases, you could be in for a crash landing.

1. 2 SOME LIMITATIONS

This book is not intended to be everything to everyone. It has some necessary limitations:

- It doesn't show you how to choose specific materials for a system, install piping, or trace out an optimum route for a pipeline. For this, have a look at *"Gravity Flow Water Supply"* (S. Arnalich 2008).

- It also doesn't show you how to use modelling software, even though it's a pretty good introduction. If you want to learn about this, try *"Epanet in Aid; How to calculate water networks by computer"* (S. Arnalich 2007) and *"Epanet in Aid; 44 progressive exercises to calculate water networks by computer"* (S. Arnalich 2008).

You can read these books for free online, or pay for downloads or printed copies here: http://www.arnalich.com/en/libros.html.

- It doesn't deal with aqueducts or canals, which also work with gravity.
- It focuses on **pipe sizing**. In gravity flow systems there are other components that require calculations, such as break-pressure tanks, reservoir and sedimentation tanks. These are dealt with in *"Gravity flow Water Supply"* (S. Arnalich 2008).

If this is your first gravity flow water system and you want to begin designing, or check if a design will work, keep on reading: this book's for you!

1. 3 HOW IT'S ORGANIZED

1. **It's progressive**. The exercises follow a logical order for designing a distribution system, and increase in difficulty. Use it as best suits you, but remember: if you do the exercises one-by-one, in sequential order, you'll probably save yourself a few headaches.

2. **It's complemented** by *"Gravity Flow Water Supply,"* where the different aspects of a project are explained in more detail, not just the design parameters. This is referred to at different points as the "Theory book" with an owl symbol. Remember you can access it free online.

3. It has **online content.** To download, just follow the links.

4. It has various **symbols** to make reading easier.

5. **There are probably a few mistakes**...no one's perfect!

   5.1 Consult the list of errata at: www.arnalich.com/dwnl/xligraxen.doc

5.2 If you come across mistakes, please let us know: publicaciones@arnalich.com.

1. 4 WORKING OUT THE UNITS AND AVOIDING MISTAKES

To put together a solid design, you'll be doing a number of very simple calculations by hand. They'll be deceptively easy to do (like working out how many days there are between 2 dates) and mistakes can always creep in, so stay sharp.

If you're disciplined when working out the units, then you'll spot most of these mistakes before you have a nervous breakdown. Have a look at this example, a conversion of m^3/h into l/s:

$$14\ m^3/h = 14\ m^3/h * 1m^3/1{,}000l * 3{,}600s/1h = 14*3{,}600/1{,}000\ m^3*m^3*s/h*l*h = 50.4\ l*m^6/h^2*s$$

$l*m^6/h^2*s$?! If you're like me and don't have a clue what this unit of flow is, something, somewhere, went a little bit wrong…

$$14\ m^3/h = 14\ m^3/h * 1{,}000l/1m^3 * 1h/3{,}600s = 14*1{,}000/3{,}600\ m^3*l*h/h*m^3*s = 3.88\ l/s$$

Note that the 2 answers are really quite different.

NOTE: Multiplying by 1h/3600s is the same as multiplying 1/1, as 1 hour and 3600 seconds are the same thing. If it's easier, think of this: "1 hour has 3600 seconds." The result is a simple change of units.

Do what I say and not what I do! To make reading the exercises easier, I sometimes leave out the units.

2. Piping

2. 1 HYDRAULICS FOR BELIEVERS

The theory behind water systems can be a little intimidating, making the learning process difficult and frustrating. In due course, you'll have time to understand the theory in more depth, once you're familiar with the basic concepts. In the meantime, you just need to have faith in the following 7 basic principles:

1. When water *isn't* flowing in a pipe, the pressure is simply the difference in height between the tap at the bottom and the surface of the water where it enters the pipe at the top. It makes no difference the route taken by the pipe on the way down.

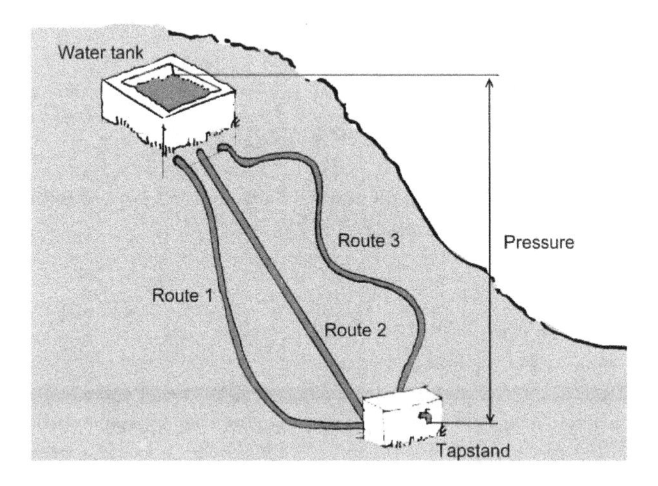

2. The pressure can be measured in metres (of column of water). 10 metres are equivalent to 1 bar or 1 kg/cm^2.

3. When water flows through a pipe, some energy is lost due to the friction of the water on the walls of the pipe.

4. Loss of pressure due to this friction can be expressed in metres per kilometre of pipeline (m/km).

5. The smaller the pipe diameter, the more pressure is lost due to frictional losses, as if the pipe was being "strangled." (Make sure you read this point carefully and understand it well[1]!)

6. The material of which a pipe is made affects the amount of frictional loss. Each kind of material has varying roughness (as well as a varying internal diameter.)

7. A system will work if the pressure is between 10 and 30 metres at the taps, and as long as the pressure at any point along the way is no less than 10 metres.

Have a look again at this last point:

> **"A system will work if the pressure is between 10 and 30 metres at the taps, and as long as the pressure at any point along the way is no less than 10 metres."**

Designing a system is all about choosing pipes of the right diameter to make sure you meet that requirement. By selecting different pipe diameters you can control the amount of pressure in the system. That way, users aren't exasperated by a miserable flow coming out of a tap (due to insufficient pressure), nor are they showered by a high-power water jet (due to excessive pressure).

2. 2 USING THE FRICTION LOSS TABLES

The simplest way of finding out the frictional loss of a given pipe at a given flow rate is using the friction loss tables. The process is simple:

1. Find the table for the given pipe material, nominal pressure and diameter. For example, PVC 90mm at 10 bar (PN10).

---

[1] *Don't get confused by the fact that if you squeeze a hose the water comes out faster. If someone puts their hands around your neck and squeezes hard you get less blood to the head and not more! This is one of the most frequent misunderstandings in cooperation projects.*

2. Look for the required flow rate and read off the head loss. For example, a flow rate of 1.25 l/s produces a 1 m/km head loss.

PVC 90 - ID 81.4mm- PN 10		
J (m/km)	Q (l/s)	v (m/s)
0.50	0.841	0.16
0.60	0.933	0.18
0.70	1.019	0.20
0.80	1.100	0.21
0.90	1.177	0.23
1.00	1.250	0.24
1.10	1.319	0.25

Pressure loss, frictional loss, head loss, hydraulic gradient or J are all the same thing.

In Appendix A & B you´ll find the generic tables. Use these if you can´t find more precise data, or if you´re still not sure where you´ll be buying your pipe.

Two important things if you´re dealing with a real-life project:

- Head losses vary a little from one manufacturer to another. If a manufacturer provides you with reliable data, use theirs.

- Pipe is specified commercially with the internal diameter for metal pipe, and the external diameter for plastic pipe (PVC or HDPE). A 25mm plastic pipe and a 25mm metal pipe have different internal diameters.

Abbreviations:

E, Energy, or head. This is the energy in the form of pressure contained in a system where there are no frictional losses.
J, Head loss factor. The energy lost as water travels through the pipe, per kilometre.
H, Head loss. The pressure lost as water travels through the pipe.
P, Pressure. The remaining pressure after frictional losses.

What is the head loss in 1 km of 110mm HDPE PN 10 pipe which carries a flow of 2 l/s?

HDPE 110 - ID 96.8mm- PN 10		
J (m/km)	Q (l/s)	v (m/s)
0.50	1.347	0.18
0.60	1.495	0.20
0.70	1.632	0.22
0.80	1.761	0.24
0.90	1.883	0.26
1.00	1.999	0.27
1.10	2.110	0.29

1. In the HDPE table for 110mm PN 10 pipe, the head loss for 1.999 l/s is: J_{110} = 1 m/km.

2. The head loss is:
   H = 1 m/km * 1 km = 1m.

What is the head loss in 5km of HDPE pipe of 63mm, PN 10, at a flow of 2 l/s?

1. In the HDPE tables, 63mm, PN 10, the head loss for 2.038 l/s is: J_{63} = 15 m/km.

2. The head loss is:

   H = 15 m/km * 5 km = 75m.

HDPE 63 - ID 55.4mm- PN 10		
J (m/km)	Q (l/s)	v (m/s)
0.50	0.293	0.12
0.60	0.326	0.14
0.70	0.357	0.15
12.00	1.799	0.75
15.00	2.038	0.85
20.00	2.393	0.99

What is the head loss in a PVC pipe of 10 bar, where the first 500m are 110mm, and the final 300m are 90mm, at a flow of 3.6 l/s?

1. Looking at the PVC tables for 10 bar, we get 2.25 m/km for 110mm pipe: J_{110} = 2.25.

2. For 3.594 l/s, J_{90}= 6.5 m/km (the head loss for 90mm pipe is 6.5 m/km).

3. The total head loss in a pipeline is the sum of the head loss in each respective reach:

   First reach 110mm: H_{110} = 0.5 km * 2.25 m/km = 1.125m
   Second reach 90mm: D_{90} =0.3 km * 6.5 m/km = 1.95m

   D_{Total} = 1.125m + 1.95m = 3.075m or 0.3075 bar.

From a tank at an elevation of 63m, a PVC pipe 2km long, of 200mm, 10 bar, is installed. This supplies a tap at an elevation of 41m. What is the pressure at the tap if there is no flow? And at a flow of 27 l/s?

1. The static pressure with the tap shut is the difference in elevation: P = 63m - 41m = 22m.
2. For 26.99 l/s, J_{200} = 4.75 m/km. The head loss is:

H = 2 km * 4.75 m/km = 9.5m.

3. The pressure with the water flowing is the static pressure less the head loss:

P = 22m – 9.5m =12.5m o 1.25 bar.

We want to supply 1.25 l/s from a spring at an elevation of 32m to a public tap stand at 15m. If the distance is 4km, what diameter of PVC 10 bar pipe would I need for the water to arrive with 13m of pressure?

Pay close attention to this exercise! It´s the first one in which you choose a pipe diameter. Once you've gotten hold of it you´re well on your way.

1. The available pressure is the difference in elevation: E = 32m - 15m = 17m.

2. The pressure that needs to be burnt off for the water to come out with 13m of pressure is:

   P = 17m – 13m = 4m

3. Those 4m are going to be burnt off over a distance of 4km, which means the head loss is:

   J = 4m / 4km = 1 m/km.

4. Look at the tables and find which diameter of PVC pipe has a head loss of 1m/km at a flow of 1.25 l/s. You´ll see that it´s 90mm.

Congratulations! You´ve just designed your first system.

PVC 90 - ID 81.4mm- PN 10		
J (m/km)	Q (l/s)	v (m/s)
0.50	0.841	0.16
0.60	0.933	0.18
0.70	1.019	0.20
0.80	1.100	0.21
0.90	1.177	0.23
1.00	1.250	0.24
1.10	1.319	0.25

2. 3 INTERPOLATING VALUES

Up to now you´ve been working with ready-made values, but it´s more likely that when you look at the tables, the figure you need is somewhere in between the ones given. To find the answer, you need to do what´s called a lineal interpolation. The generic equation looks a bit muddled, but it´s fairly simple:

$$\frac{J_x - J_{\text{inf}}}{J_{\text{sup}} - J_{\text{inf}}} = \frac{Q_x - Q_{\text{inf}}}{Q_{\text{sup}} - Q_{\text{inf}}}$$

Where:
J_x, the head loss you are looking for.
J_{inf}, J of the lower flow rate.
J_{sup}, J of the higher flow rate.
Q_x, the flow rate in question.
Q_{inf}, the lower flow rate.
Q_{sup}, the higher flow rate.

If you don´t want to learn equations or depend on this book, the geometric reasoning is very simple (although if you prefer to skip it I won´t be offended).
In 90° triangles, the distances are proportional. In other words, *a/A* must be equal to *b/B*. The distance *a* is J_x-J_{inf}, *A* is J_{sup}-J_{inf}, *b* is Q_{sup}-Q_x and *B* is Q_{sup}-Q_{inf}.

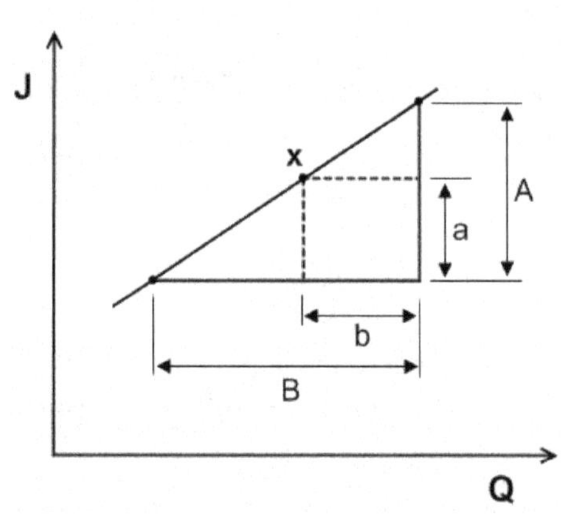

As you are going to be interpolating for even the simplest systems, draw up a spreadsheet with Excel or use this one:

www.arnalich.com/dwnl/interpolator.xls

In a PVC 10 bar, 90mm pipe, what is the head loss at 6 l/s?

1. Look at the PVC tables. There´s no value that´s sufficiently close: it´s either 5.723 l/s or 6.71 l/s.

2. Using the formula you can work out the head loss:

$$\frac{J_x - J_{\text{inf}}}{J_{\text{sup}} - J_{\text{inf}}} = \frac{Q_x - Q_{\text{inf}}}{Q_{\text{sup}} - Q_{\text{inf}}}$$

$$\frac{J_x - 15}{20 - 15} = \frac{6 - 5.723}{6.71 - 5.723}$$

J_x = 15 + (20-15) (6 – 5.723)/(6.71-5.723) = 16.4 m/km

Using the spreadsheet calculator provided, you should get the same result:

	Lower value	x	Higher value
Flow	5.723	6	6.71
J	15	**16.4**	20

2. 4 COMBINING PIPE SIZES

Imagine you want to transport 2 l/s and that you need to burn off 10m/km for the water to arrive at the right pressure. The 63mm HDPE loses 15m/km and the next pipe size along, 90mm, only loses 2.5 m/km. The pipe you need isn´t available!

The solution is to combine pipe sizes until you get the desired head loss. To work out the required pipe lengths use the following reasoning:

The head loss for x meters in pipe A together with the head loss for x meters in pipe B is the total required head loss.

Translated into an equation:

$$J_a{*}x + J_b(d-x) = D$$

Where, J_a, is the head loss for pipe A
J_b, head loss for pipe B
x, length of pipe A
d, is the total length
D, required head loss

If *x* is the length in meters of pipe *A* that needs to be installed, the length of pipe *B* is *d-x*.

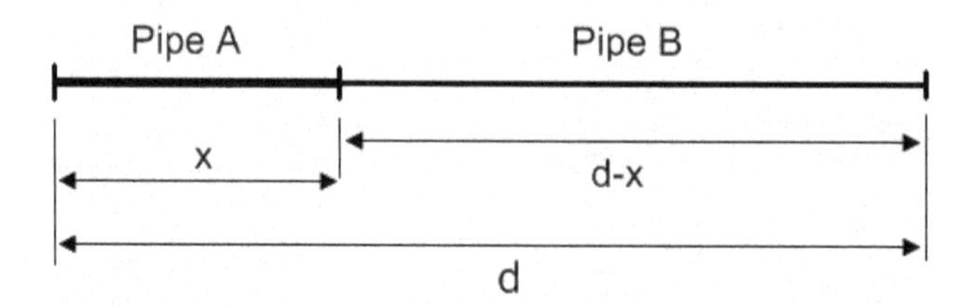

It´s very important to remember that **the larger pipe diameter is almost always installed first,** such that the water travels through progressively smaller pipe sizes. In general it makes no sense to use smaller pipes followed by larger ones. If new connections are made in the future or the demand increases, a small diameter pipe placed too soon in the pipeline will strangle the rest of the system.

Which HDPE pipe would you install for 20 meters of head loss over 2 kilometres at a flow of 2 l/s?

1. Look at the head loss values for the pipes:

   J_{63}= 15 m/km. For 2 km you lose 15m/km * 2km = 30m. Too much!
   J_{90}= 2.5 m/km. For 2 km you lose 2.5m/km * 2km = 5m. Too little!

2. As you may have suspected, the solution lies in combining pipe sizes:

$$J_a * x + J_b(d-x) = D$$

$$J_{63} * x + J_{90}(d-x) = D$$

$$15\ m/km * x + 2.5\ m/km\ (2-x) = 20m$$

$$15x + 5 - 2.5x = 20$$

$$12.5x = 15 \quad \rightarrow \quad x = 1.2\ km\ of\ pipe\ A$$

$$d-x = 2\ km - 1.2\ km = 0.8\ km\ of\ pipe\ B$$

When you get more familiar with the process, you´ll end up using trial and error with the formulas in the spreadsheet, rather than using these equations. Either way, take it one step at a time.

2. 5 ADDING THE TOPOGRAPHIC PROFILE

Up until now we´ve ignored the path taken by the pipe. It has simply been a matter of the start and end points. If you feel comfortable using the tables we can complicate things a little and look at the topographic profile taken by the pipe line.

Here, the important thing is the last part of the last principle we looked at. Take a moment to look it over again and refresh your memory.

> "[The pressure]… at no point along the pipeline should be less than 10 meters."

Imagine you have a profile similar to this:

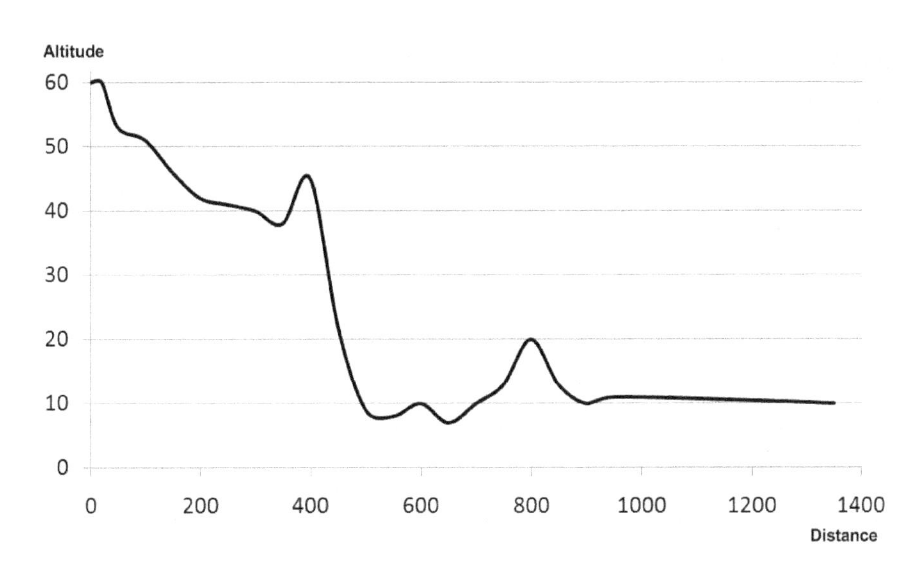

For practical purposes, if you manage to get over the main obstacles by over 10 meters, you don´t have to worry about the rest of the pipe line:

In this profile, there are 2 critical points, the high points *A* and *B* at 400 and 800m. We´re going to deal with them in the following exercise.

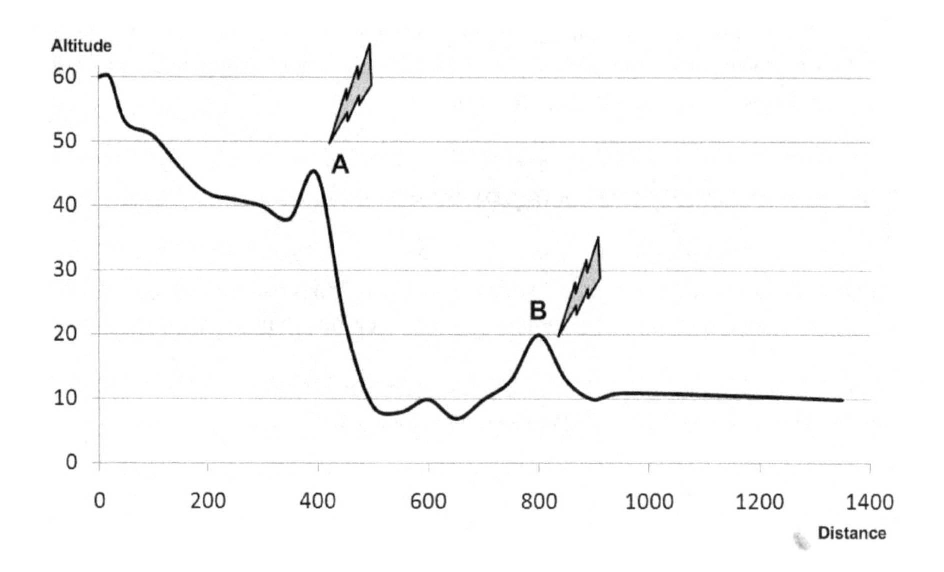

Which HDPE pipe would you use to get to 1,400 meters with 2 bar of residual pressure at a flow of 4 l/s?

1. Look for the critical points. Note that *B* is not really that critical, since the elevation 10 meters above is 30m, and the elevation at which we want to deliver the water is 10m + 20m = 30m (remember, 2 bar is equivalent to 20m). The exit point and *B* are therefore really at the same height.

2. Let´s work out the first reach of pipe to get to *A* with at least 10m of pressure. In other words, with 46m + 10m. The maximum head loss will be:

   J_{max}= (60-56m) / 0.4 km = 10 m/km.

3. In the tables, J_{90}= 9 m/km. The head loss in that reach is:

   D= 0.4 km * 9 m/km = 3.6m

4. For the second reach, the start point is 60m – 3.6m = 56.4m of energy. Since the end point is 30m, the required head loss is:

   J= (56.4- 30)m / (1.4-0.4)km = 26.4 m/km

5. In the tables, J_{90}= 9 m/km. To find the value of J_{63} we need to interpolate:

	Lower value	x	Higher value
Flow	3.752	4	4.396
J	45	**50.78**	60

6. We need to combine pipe sizes in the last 1000 meters:

$$J_{63} * x + J_{90}(d-x) = D$$

$$50.78 \text{ m/km} * x + 9 \text{ m/km } (1-x) = 26.4\text{m}$$

$$50.78x + 9 - 9x = 26.4$$

$$41.78x = 17.4 \quad \rightarrow \quad x = 0.416 \text{ km of 63mm pipe}$$

$$d-x = 1 \text{ km} - 0.416 \text{ km} = 0.584 \text{ km of 90mm pipe}$$

So we´ll use 400m + 584m = 984m of 90mm HDPE, PN 10, followed by 416m of 63mm HDPE, PN 10 (in that order!)

If you are tempted to run off and design pipelines now, be patient and finish this chapter. There are still some things that have deliberately not been dealt with to keep the exercises simple, and which could scupper your efforts so far.

2. 6 VISUALISING PRESSURE AND ENERGY

Meters of energy? Meters of pressure?

You´ll soon discover the reason we´re using such unorthodox units: they let us visualise everything at the same scale.

In this section, you´ll learn how to *see* your calculations. You´ll see the profile which shows the energy of the water in the pipe, together with the topographic profile. The vertical distance between these 2 profiles indicates the pressure.

This is the graph of exercise 8 which you´ve just done:

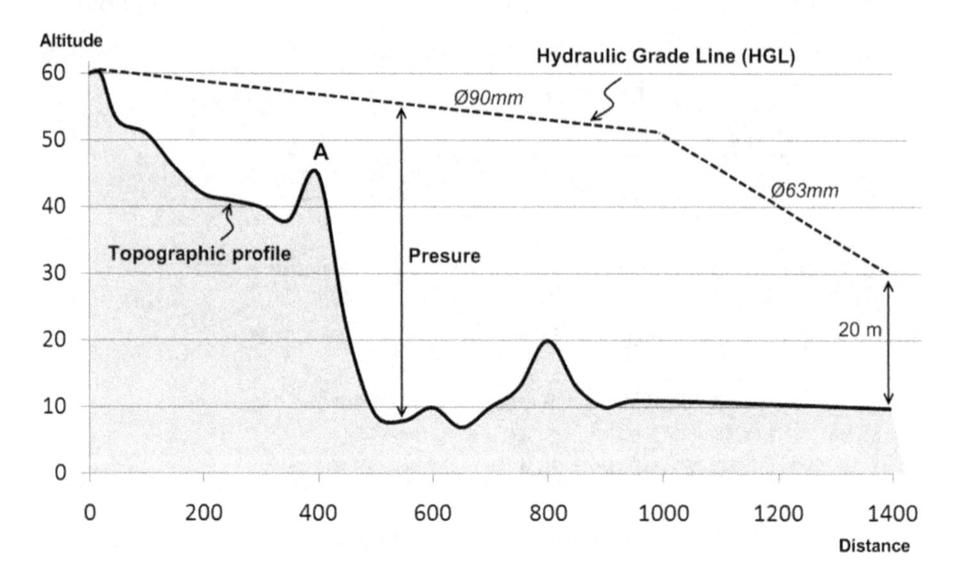

The **hydraulic grade line (HGL** from now on) represents the energy the water has. It started with 60 *meters of energy*, all of which are due to the elevation. The pressure at any particular point is the difference between the hydraulic grade line and the elevation of the terrain itself.

As energy is lost due to friction in the pipeline, the HGL begins to incline. This incline is the J you have calculating so far. In the first reach of 90mm pipe, up until it reduces at 984m, the frictional loss was 9 m/km. After that, it inclines further with the reduction to 63mm, reflecting the head loss of 50.78 m/km.

Now is a good moment to read sections 1.4 and 1.5 of the theory book. Section 1.5, *The hang glider analogy,* presents you with a more intuitive way of understanding the calculations. Don´t skip this bit! You can carry on with the *Initiation flight* in section 1.6, which is very similar to exercise 8.

Take note: it´s easy to make the mistake in thinking that the HGL is where the pipe actually goes. Remember, the pipe follows the terrain profile!

The path taken by a pipe of a given length between one point and the next, does not affect the pressure at either of those points. In other words, the outlet pressure of a pipe is the same whether it´s rolled up, in a dead straight line, or laid out in whichever way possible.

Design a PVC mainline to supply a reservoir tank with a constant flow of 5 l/s, over the supplied profile, with a residual head of 1 bar at the outlet.

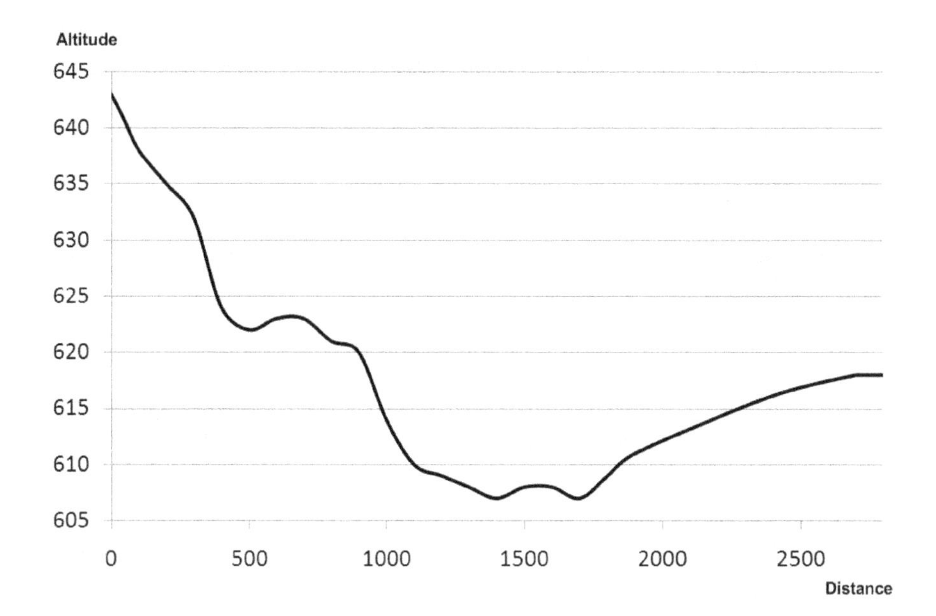

1. Define the outlet point and look for the critical areas. The outlet is the elevation, 618m, together with the required residual pressure, 1 bar: 618m + 10m = 628m.

Pay close attention to the scales of the topographic survey. At first glance, the profile can give you a false idea that there´s a big drop and that there won´t be any critical points. If you adjust the vertical scale, you can see the situation more clearly:

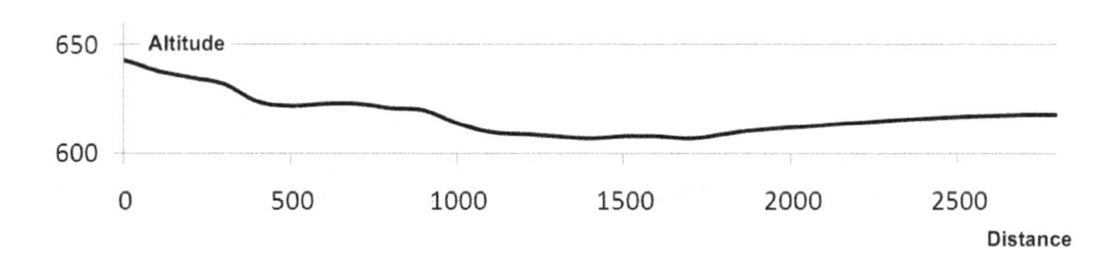

You can understand the need for precise calculations once you see the profile at the correct scale…in reality, it´s very flat!

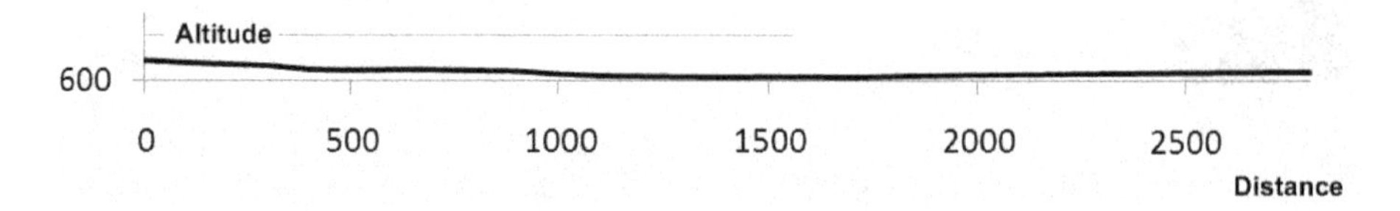

Returning to the enlarged profile, there are 2 obstacles before reaching the outlet. Point *A* because it´s so close to the inlet. Point *B* is a small obstacle. Point *C* is the outlet with 1 bar of pressure.

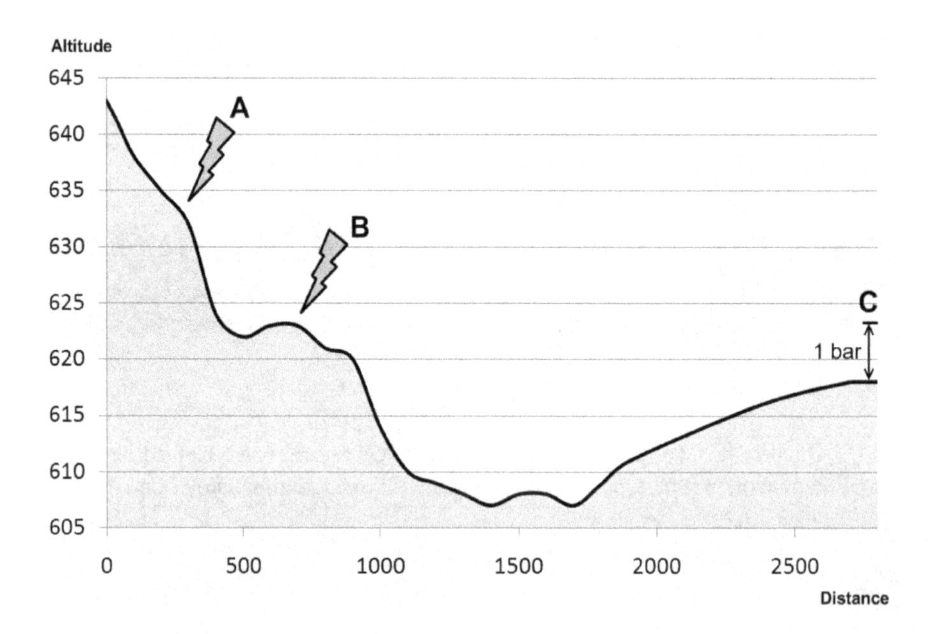

One way of doing it is to aim straight for point *C* and check if the HGL gets dangerously close to the other points:

2. The maximum head loss is: 643m – 618m - 10m = 15m. Seeing you can lose 15m in 2.8 km, the HGL is: J_{max} = 15m / 2.8 km = 5.36 m/km.

You can work out the height of the HGL at the critical points by calculating the grade line as you move down along the pipe, i.e. multiplying it by the distance at each point:

3. The first is at an elevation of 632m and a cumulative distance or **chainage** of 300m. The head loss will be: D = 0.3 km * 5.36 m/km = 1.6m

The pressure at this point will be the total available pressure, less the head loss, and less the elevation of the terrain at that point: P = 643m - 1.6m - 632m = 9.4m.

Despite being less than 10 meters, this is not a big problem, since the system can´t build up the necessary pressure straight away: it needs a certain distance and drop before it can pressurise sufficiently. This means *A* isn´t really a critical point.

4. The second is at a distance of 600m and at an elevation of 623m. Repeat the process:

D = 0.6 km * 5.36 m/km = 3.22m
P = 643m – 3.22m - 623m = 16.78m. Not critical.

5. Seeing as there are no obstacles, you can aim straight for the outlet. To find out which PVC pipes are needed, look for head losses of around 5.36 m/km for 5 l/s:

J_{110} = 4 m/km at a flow of 4.97 l/s
J_{90} = 12 m/km at a flow of 5.057 l/s

110mm pipe is sufficiently close to avoid having to combine pipe sizes. The pressure at the outlet would be:

P = 643m – 618m – 2.8 km * 4m/km =13.8m.

If it was essential to have exactly 10 meters of residual pressure, you´d need a combination of pipe sizes.

Size an HDPE mainline to feed a distribution system with a peak flow of 8 l/s, with a residual pressure of between 1 and 3 bar, using the profile below.

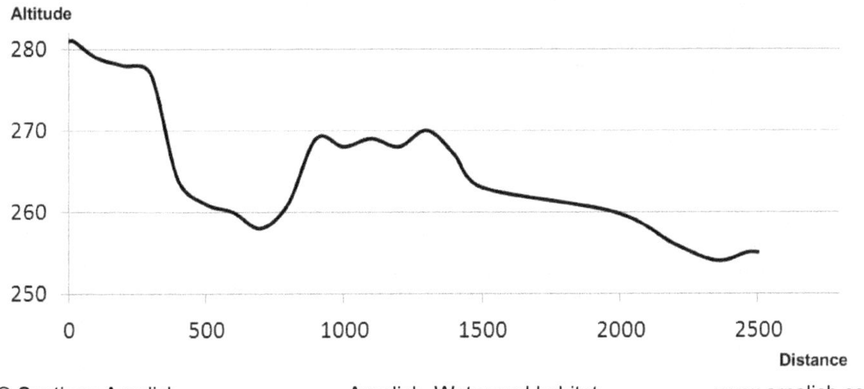

1. In this example, it´s clear there´s a critical point at 1,300m. The HGL will also be close to ground level at 300m. At this point, it´s not possible to have 10 meters of pressure due to the lack of available height.

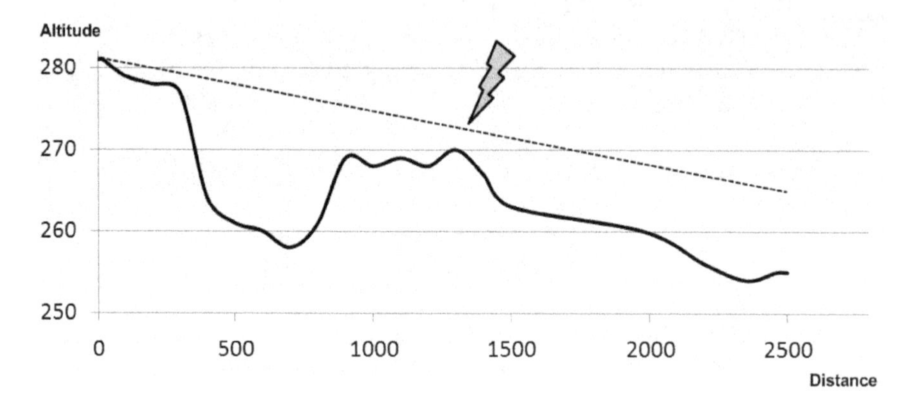

2. To reach the point at 1300m with 10 meters of pressure, the maximum head loss is:

$$J_{max} = (281m - 270m + 10m) / 1.3km = 0.77m/km$$

3. Using the HDPE tables, J_{200} = 0.7 m/km for 8.221 l/s. To find the corresponding value at 8 l/s you can interpolate:

	Lower value	**x**	**Higher value**
Flow	7.539	8	8.221
J	0.6	**0.67**	0.7

Note that for such low head loss values you don´t gain much in terms of precision, only 0.007 bar in this case.

4. The pressure at the 1300m point is: P = 281m - 270m- 1.3km * 0.67 m/km = 10.13m.

5. The HGL is almost flat in the first reach, which also means that the first 300 meters have most of the available pressure.

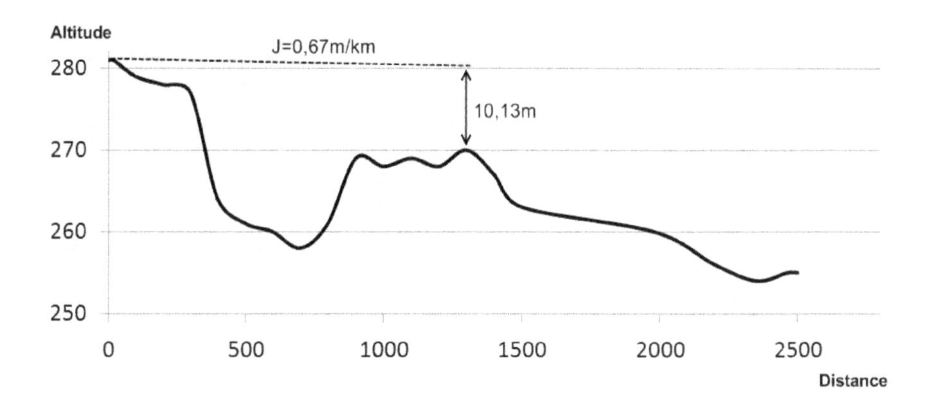

6. The second reach begins at 280m, with 13m to reach 265m (255 + 10m). The available energy is 15.13m (280.13m – 265m), over a distance of 1,200m:

$$J_{max} = 15.13m / 1.2 km = 12.61 m/km$$

7. In the HDPE tables for 110mm and 10 bar, at a flow of 8.04 l/s, the head loss is 12m/km: J_{110}= 12 m/km.

8. As the previous reach left at 280.13m, the pressure at the outlet is:

$$280.13 m – 255 m – 1.2 km * 12 m/km = 10.73 m \quad (1.07 bar)$$

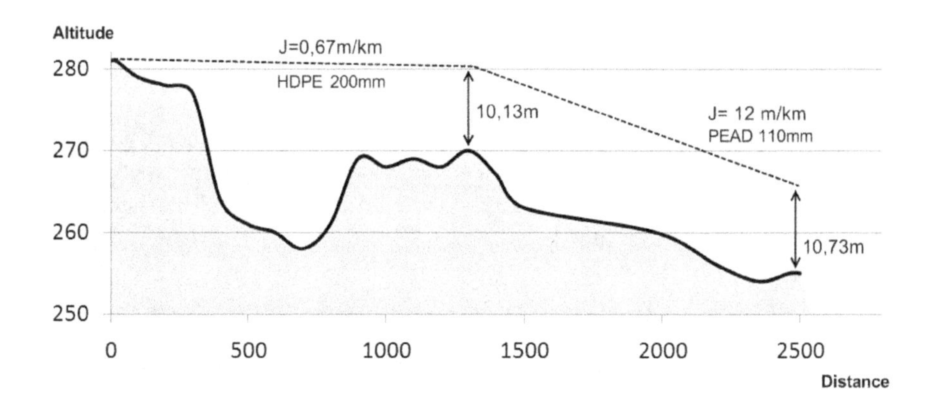

Ten meters is the minimum pressure and 10.73 isn´t much more than that. In the real world, it´s better to leave yourself a bigger margin. To do so, would mean lengthening the 200mm pipe a little. Work out on your own how much pressure you´d gain if it was 200 meters longer. (Answer: 13m)

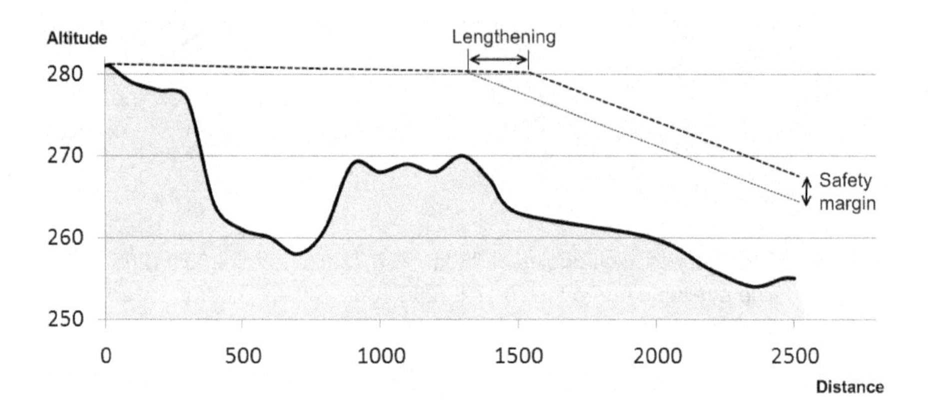

2. 7 MAXIMUM WORKING PRESSURE IN PIPING

Up to now, you´ve been looking at the peaks to avoid low pressures. But the valleys can also cause problems if the pressure is too much for the pipe. Under normal working conditions, try and make sure the pipe operates at **no more than 80%** of its specified pressure limit. That means:

For pipes of: PN10, 10 bar * 0.8 → 8 bar maximum pressure
PN16, 16 bar * 0.8 → 12.8 bar maximum pressure

There´ll be maximum pressure in the pipe when the water is not flowing and no pressure is lost due to friction. This means to calculate maximum pressures you don´t need to deal with flow or head loss. Here the HGL is horizontal (also referred to as a point of static pressure) and all you need to do is look at the difference in height between the HGL and the low points.

For example, in exercise 10 the maximum pressure will be in the lowest valley:

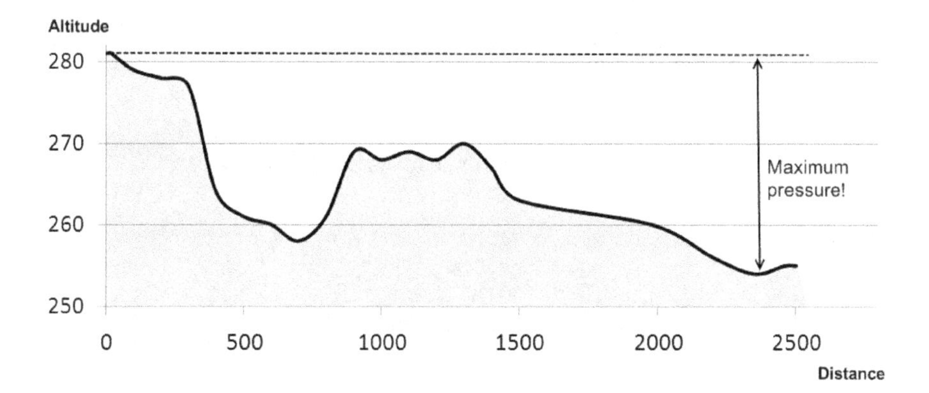

The maximum pressure is the difference between the elevations: 281m – 253.5m = 27.5 m. Since 2.75 bar is less than 8 bar, you can install PN10 pipe.

From a spring at an elevation of 345 meters, you want to feed a reservoir tank at an elevation of 324m with a continuous flow of 3 l/s. The distance to be covered is 3.75km. Select the right piping for the supplied profile, taking into account that the pipe will be anchored but not buried:

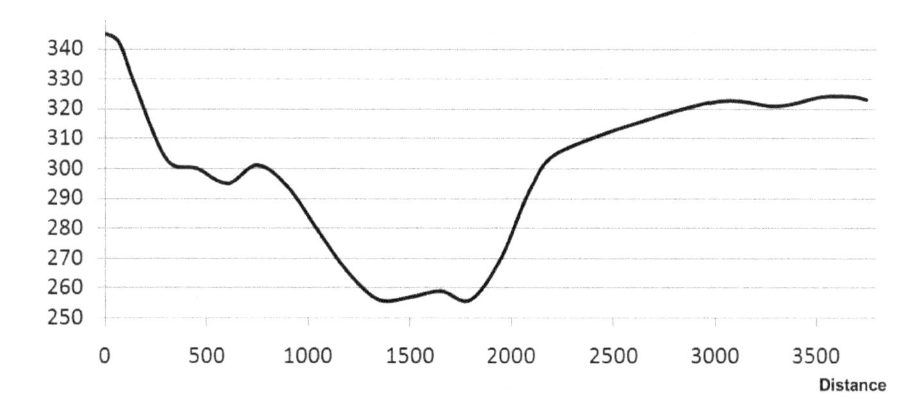

1. The difference between the highest point, 345m, and the lowest, 255m, is 90m. When the water is at rest you'll have 90m of pressure, which exceeds by 10 meters the 80% pressure limit for 10 bar pipe. At some points in the pipeline you´ll need to install PN16 pipe.

2. To work out where, subtract the maximum pressure, 80m, from the highest elevation:

$$345m - 80m = 265\ m$$

Any point below 265m means you need to install 16 bar pipe.

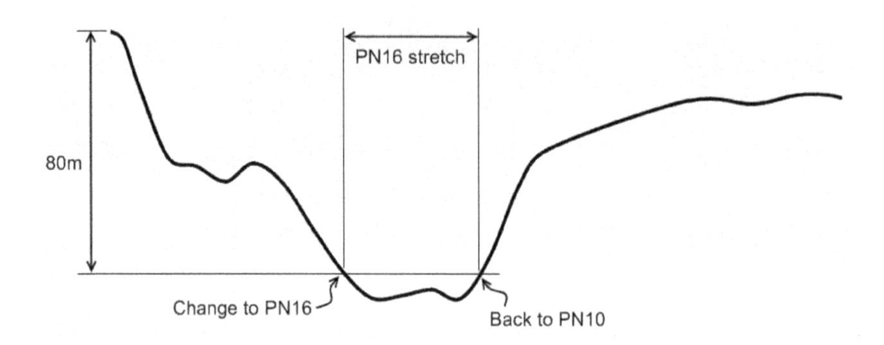

PN 16 pipe has a higher frictional loss than PN 10, which is why the tables have differing values for each pressure rating. Knowing beforehand which sections require PN 16 will save you a lot of work.

3. Use the same method you´ve been using so far to work out the pipe diameters, paying special attention to the points where the pipe diameters change. As there are no obstacles, you can aim straight for the outlet point at 324 meters + the residual pressure.

The outlet is freely flowing into a reservoir tank. To avoid wear of the parts and minimise repairs and replacements, and also to avoid a situation where the float valve won´t close due to excessive pressure, it´s best to arrive with the minimum pressure of 10m.

4. The outlet point is 324m + 10m = 334 m. If the starting point is at 345m, maximum head loss is 11 metres: 345m – 334m = 11m. The resulting head loss is:

$$J = 11m / 3.75\ km = 2.93\ m/km$$

PVC degrades in the sun, so you can´t install it without burying it. That means you have to look at the HDPE tables.

5. You´ll see that at 2.957 l/s, J_{110}= 2 m/km. This value is close enough and leaves some margin.

6. The distance up until the transition to PN 16 is 1,200 m. So the head loss in the first reach is:

$$D = 2 \text{ m/km} * 1.2 \text{ km} = 2.4\text{m}$$

7. The PN 16 section is 600m long. In the HDPE tables 110mm, PN 16, at 3.053 l/s, J_{110} = 3 m/km. The head loss in the second reach is:

$$D = 3 \text{ m/km} * 0.6 \text{ km} = 1.8\text{m}.$$

8. Of the 11m of head we could lose, and after the head loss in the previous 2 reaches, we are left with: 11m – 2.4m – 1.8m = 6.8m.

   The length of the final reach is 3,750m – 1,200m – 600m = 1,950m.

   The required head loss is: J= 6.8m / 1.95 km = 3.49 m/km.

9. As there´s no PN 10 pipe which gives us that exact head loss, we need a combination of 90mm and 110mm pipe. J_{90}= 5.5 m/km y J_{110}= 2 m/km.

$$J_a*x + J_b(d-x) = D$$

$$J_{90}*x + J_{110}(d-x) = D$$

$$5.5 \text{ m/km} * x + 2 \text{ m/km} (1.95-x) = 6.8\text{m}$$

$$5.5x + 3.9 - 2x = 6.8$$

$$3.5x = 2.9 \rightarrow x = 0.829 \text{ km of 90mm pipe.}$$

$$d-x = 1.95 \text{ km} - 0.829 \text{ km} = 1.121 \text{ km of 110mm pipe.}$$

To build this system the following pipe is needed:

1,200m + 1,121m = 2,321m of 110mm HDPE PN10
600m of 110mm HDPE PN16
829m of 90mm HDPE PN10

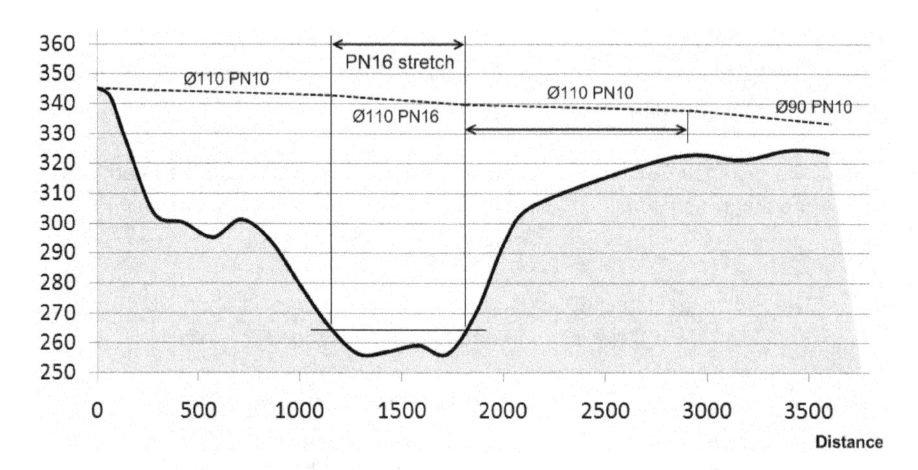

2. 8 BRANCH LINES

In this section you'll learn how to design distribution systems with branch lines, which are the most common. The procedure is exactly the same, you just have to take into account the flow in each respective pipe and calculate the branch lines one by one.

In a system with just one branch line and 2 final consumer nodes of 3 and 2 l/s, the mainline carries the total flow, and then divides into each branch line:

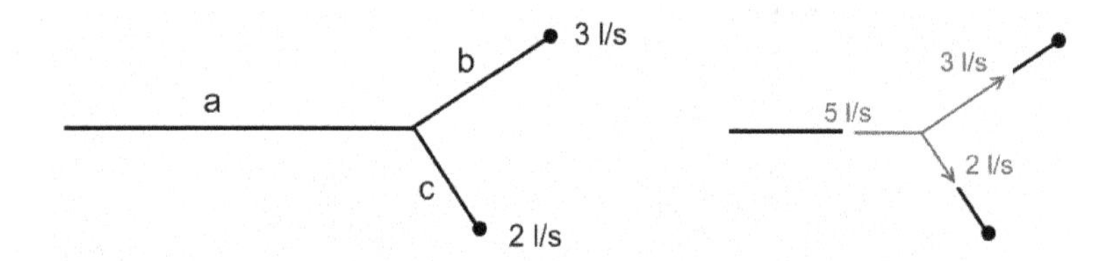

The procedure is the same for adding another branch line. Pipes *d*, *c* and *e* carry the final consumer demands of 1, 3 and 2 l/s respectively. *b* carries the flow for *d* and *c*: 1+3 l/s = 4 l/s, and *a* the sum of *b* and *e*, 4+2= 6 l/s.

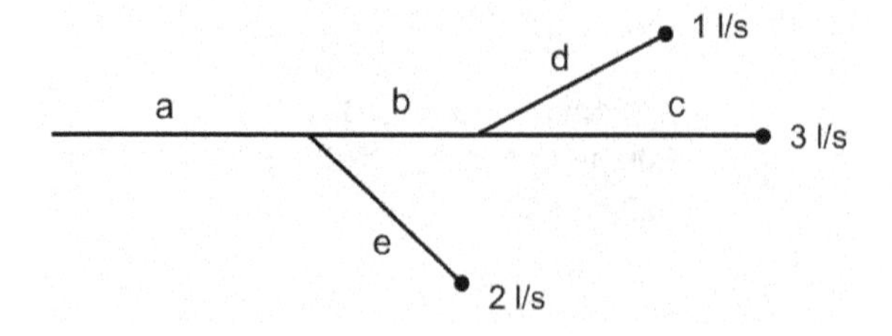

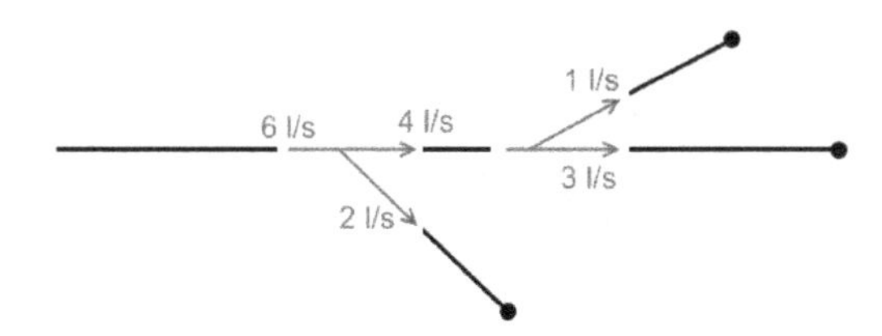

Size the PVC pipes needed for the following system, fed by a spring at 32m, with 2 public tap stands at 7m and 0m.

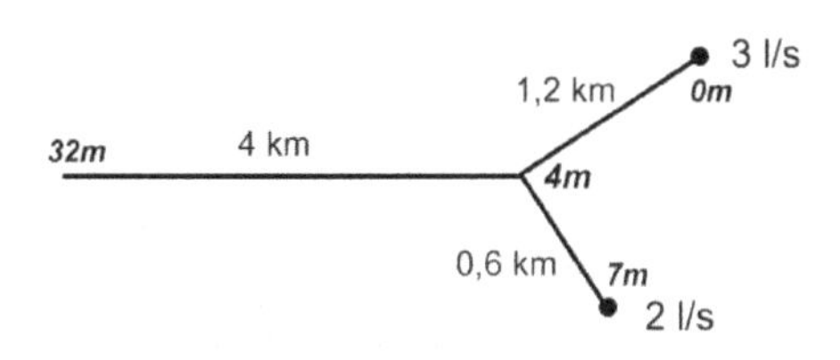

1. The first pipe carries the flow for the entire system, 3+2= 5 l/s. The other two only have to transport the final demand at each node. We´ll call *a* the 4km pipe, *b* the 1.2km pipe, and *c* the final pipe.

2. There is no point in the system where the pressure is more than 8 bar, so all the pipes can be PN 10.

3. The energy in the branch line must be enough to allow the water to reach each point. Choosing the pipe is a skill you´ll develop with practice. In the meantime, use trial and error.

4. Seeing as the highest point is 7m, it makes sense to try and reach the branch point with 20m. This leaves 3m of energy to burn off in pipe *c*, if there´s 1 bar residual pressure: 20m – 7m – 10m = 3m.

5. To arrive with 20m, the head loss for *a* is:

$$J = (32m - 20m) / 4\ km = 3\ m/km$$

6. Looking at the PVC tables for 5 l/s, J_{110}= 4 m/km and J_{160}= 0,7 m/km. We can´t install 4km of 110mm otherwise we won´t reach the branch lines with enough pressure. On the other hand, if we install 4km of 160mm the project may end up being too expensive.

If you´re at the design stage and there are likely to be system enlargements, it´s best to use a larger pipe. If the project is tight on funds, you´ll have to combine pipe sizes, as you´ve seen in the previous exercises.

7. Since funds are almost always in short supply, we combine pipe sizes:

$$J_{110}*x + J_{160}(d-x) = D$$

$$4 \text{ m/km} * x + 0.7 \text{ m/km} (4 - x) = (32-20)\text{m}$$

$$3.3x + 2.8 = 12; \quad 3.3x = 9.2 \rightarrow x = 2.79 \text{ km of 110mm pipe}$$

$$d-x = 4 \text{ km} - 2.79 \text{ km} = 1.21 \text{ km of 160mm pipe.}$$

The first reach will be 1.21km of 160mm, followed by 2.79km of 110 mm.

8. In pipe *b* the flow is 3 l/s over a distance of 1.2km. Starting with 20m of energy at the branch point, the head loss to arrive with at least 1 bar is:

$$J_{max} = (20\text{m} - 0\text{m} - 10\text{m}) / 1.2 \text{ km} = 8.3 \text{ m/km.}$$

9. Looking at the tables for 3 l/s, J_{63} is a value between 20 and 30 m/km (too much). J_{90} = 4.75 m/km. Over a distance of 1.2km, the head loss is D = 1.2 km * 4.75 m/km = 5.7m. If we start with 20m, the pressure at the outlet of pipe *b* will be:

P= 20m – 0m – 5.7m = 14.3m or 1.43 bar, which give us some margin above the minimum.

10. Pipe *c* distributes 2 l/s over 0.6km, so:

$$J_{max} = (20\text{m} - 7\text{m} - 10\text{m}) / 0.6 \text{ km} = 5 \text{ m/km.}$$

$$J_{90} \approx 2.5 \text{ m/km}$$

$$P = 20\text{m} - 7\text{m} - 2.5 \text{ m/km} * 0.6 \text{ km} = 11.5 \text{ m or } 1.15 \text{ bar.}$$

The final result is:

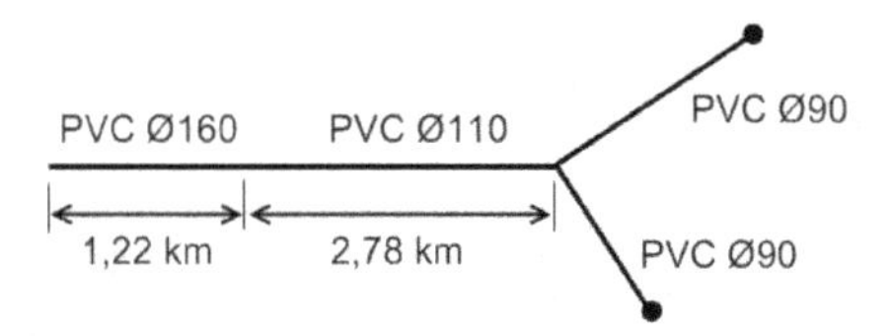

Size the PVC pipes needed for the following system to work, from a reservoir tank at 102m to 3 further tanks in three villages at elevations of 71m, 81m, and 12m respectively.

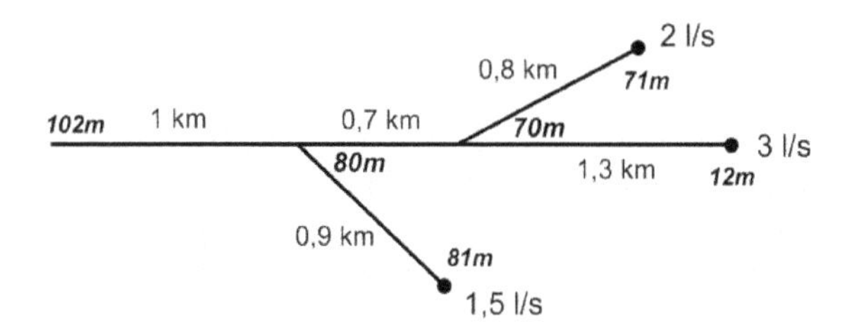

1. Since we´re dealing with reservoir tanks, we´re going to try and arrive with the minimum pressure, 1 bar, so as to avoid wear in the float valves and save on piping.

2. The flows are the following:

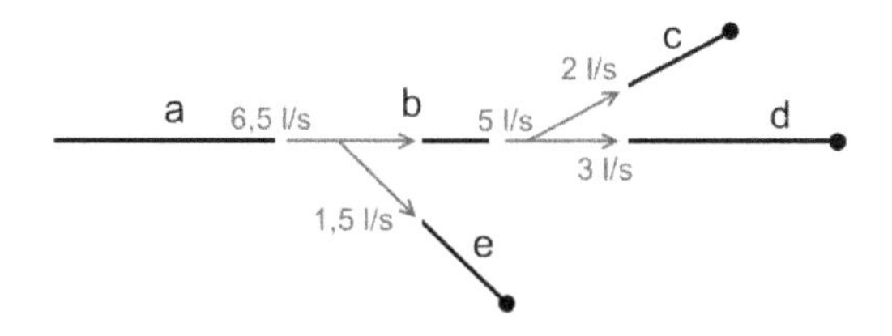

3. The pipe *d* will need PN16 at some point: 102 – 12 = 90m > 80m. Seeing as there´s no topographic profile, and to keep things simple in this exercise, we´ll assume it´s the final 200m that needs PN 16.

4. If we aim to get to the first junction with 95m:

   J= (102m -95m) / 1km = 7 m/km

$J_{110,\ 6.5\ l/s}$ = 6.5 m/km

E = 102m – 6.5 m/km * 1 km = 95.5m → Pipe *a* 110mm.

5. Pipe *e*:

J_{max} = (95.5m – 81m -10m) / 0.9 km = 5 m/km

$J_{63,\ 1.5\ l/s}$ = 7.5 m/km $J_{90,\ 1.5\ l/s}$ = 1.4 m/km (combination of pipes needed)

J_{63}*x + J_{90} (d-x) = D

7.5 m/km * x + 1.4 m/km (0.9 -x) = (95.5 -10 -81)m

6.1x + 1.26 = 4.5m

6.1x = 3.24 → x = 531m of 63mm pipe

d-x = 0.9 km – 0.531 km = 369 m of 90mm pipe.

6. Pipe *b*, aiming for 85m at the second branch point:

J= (95.5m -85m) / 0.7km = 15 m/km

$J_{90,\ 5\ l/s}$ = 12 m/km

E = 95.5m – 12 m/km * 0.7 km = 87.1 m → Pipe *a* 90mm.

7. Pipe *c*:

J_{max} = (87.1m – 71m -10m) / 0.8 km = 7.625 m/km

$J_{63,\ 2\ l/s}$ = 12 m/km $J_{90,\ 2\ l/s}$ = 2.25 m/km

J_{63}*x + J_{90} (d-x) = D

12 m/km * x + 2.25 m/km (0.8 -x) = 87.1m – 71m -10m

9.75x + 1.8m = 6.1m

$9.75x = 4.3 \rightarrow x = 441$m of 63mm pipe

d-x = 0.8 km – 0.441 km = 359 m of 90mm pipe.

8. As you saw in point 3, the pipe *d* has 2 reaches, one of 1.1km PN 10, and the other of 0.2km PN 16. Both have to lose head for the pressure to be between 1 and 3 bar:

J_{max} = (88.5m – 12m - 10m) / 1.3 km = 51.15 m/km → 1 bar
J_{max} = (88.5m – 12m - 30m) / 1.3 km = 35.76 m/km → 3 bar

Interpolating, $J_{63,\ 3\ l/s\ ,\ PN10}$ = 26.39 m/km

	Lower value	x	Higher value
Flow	2.583	3	3.236
J	20	**26.39**	30

Ideally you´d combine pipe sizes in which the first reach is PN10 and the second is PN16. Over 1.1km, the head loss will be 26.39 m/km * 1.1 km = 29.03m.

The remaining pressure to burn off at the end of 1.1km is:

P = (88.5m – 12m – 29.03m) = 47.47m

The 200m reach needs a head loss which leaves the pressure at a maximum of 3 bar:

$J_{min,\ PN16}$ = (47.47m - 30m) / 0.2 km = 87.35 m/km

Woops, there are no values high enough in the tables! This is a rare occurrence, but sometimes comes up in very rough terrain. In these cases, you can calculate the values you´re looking for using the Hazen-Williams formula, found at the end of Appendix B, using a friction coefficient of C-140 for PVC and HDPE and C-120 for galvanized iron. Be careful with these values if you´re using pipe of less than 50mm and velocities higher than 3 m/s.

9. The value for a 40mm pipe (careful: interior diameter is 36.2mm) is 252 m/km according to the Hazen-Williams formula: too much.

In these cases, instead of installing a combination of pipes, where one reach may be less than 100m, it´s easier to install a pressure reducing valve, which strangles the flow to dissipate the excess pressure. Another alternative is to pass the flow through a very small hole:

Fig. Frictional hole diffuser, Jmeijmeh, Lebanon.

If the pipe is small enough, less than 25mm, and there aren´t many points, you could also install pressure reducing valves (PRV), which are cheap and easy to find at those diameters:

Fig. Pressure reducing valve (PRV), Velrub, Azerbaijan.

In this case, the system would look like this:

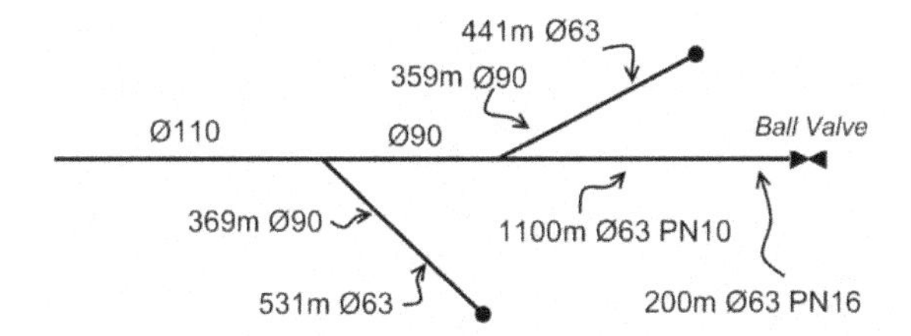

*Next you have a **confirmation exercise**, with fewer explanations and more realistic topographic profiles. It´s basically like the previous exercise, but with the profiles. If you follow it through and can do it on your own, congratulatons! The rest of the book will be all downhill. If that´s not the case, carefully take the time to go back over the previous pages.*

Calculate the proposed system for HDPE with a residual pressure at the taps of 1.5 to 3 bar, with the following topographic data:

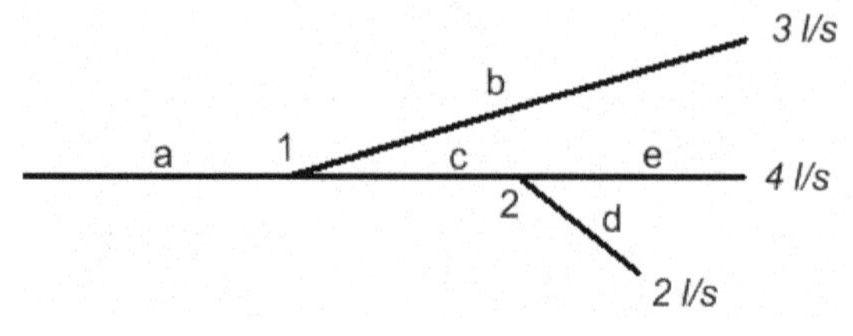

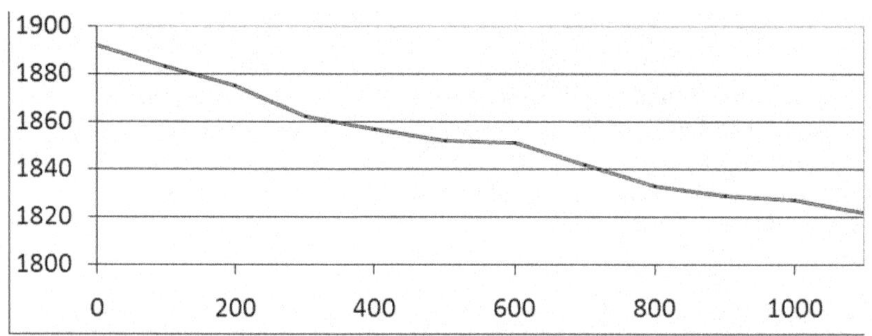

Elevation	1892	1883	1875	1862	1857	1852	1851	1842	1833	1829	1827	1822
Distance	0	100	100	100	100	100	100	100	100	100	100	100
Chainage	0	100	200	300	400	500	600	700	800	900	1000	1100
						1		**2**				**F2 (4 l/s)**

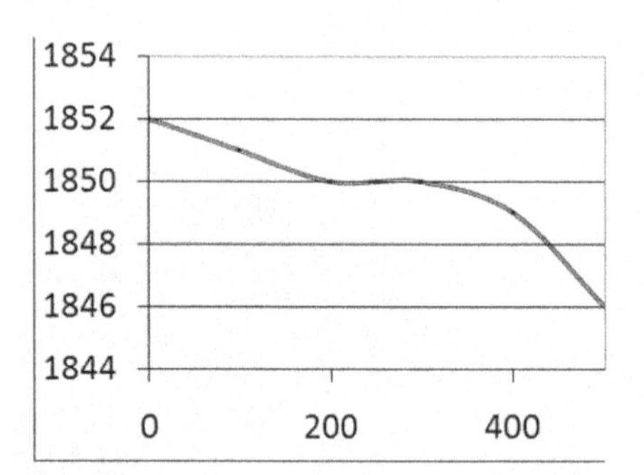

Elevation	1852	1851	1850	1850	1849	1846
Distance	0	100	100	100	100	100
Chainage	0	100	200	300	400	500
	1					**F1 (3 l/s)**

1850
1848
1846
1844
1842
1840
0 100 200

Elevation	1842	1845	1850
Distance	0	100	100
Chainage	0	100	200
	2		**F3 (2 l/s)**

To avoid aiming for very low pressures, take 15m as the minimum design pressure. If you then get values between 15m and the absolute minimum of 10m, you can decide whether it´s acceptable or not. The idea is to leave more than the minimum pressure for unknowns. If it is impractical to have it above 15m at some point, then 10m is acceptable.

REACH A

The pipe *a* has to transport 9 l/s. At point 1, the elevation is 1,852m, very close to that of the sources F1 (1,846m) and F3 (1,850m). We need enough residual head for the branch lines. With 10m we´ll be left short. We can tentatively aim for 25m of pressure, which allows for a head loss of:

1,852m + 25m of pressure = 1,877m

(1,892m-1877m)/0.500km = 30m/km

In the tables, there´s no value close enough to 9 l/s for 90mm pipe. Use lineal interpolation to find the answer:

$$\frac{J_x - J_{\inf}}{J_{\sup} - J_{\inf}} = \frac{Q_x - Q_{\inf}}{Q_{\sup} - Q_{\inf}}$$

Where:
- J_x, the head loss you are looking for.
- J_{inf}, J of the lower flow rate.
- J_{sup}, J of the higher flow rate.
- Q_x, the flow rate in question.
- Q_{inf}, the lower flow rate.
- Q_{sup}, the higher flow rate.

$$\frac{J_x - 30}{45 - 30} = \frac{9 - 7.798}{9.74 - 7.798}$$ → J_x = 39.28m/km

The dissipated energy: 0.5km * 39.28m/km = 19.64m

The residual pressure at *1* is: 1,892m- 1,852m – 19.64m = 20.36m

20.36 meters is the closest value to the 25 mentioned at the start.

REACH C

The pipe *c* has to transport 6 l/s in total, over a distance of 200m. Keeping the pressure of point 2 at 20m, the water will have enough pressure to climb the branch line *d*:

J_{max}= (20.36m + 1852m – 1842m -20m)/0.2 km = 51.8m/km

No pipe is close to this value. However, since the distance is small, you don´t have to get it exactly right on 20m (22m or 24m are equally valid), and we can use 90mm pipe. For 6.2 l/s, J= 20 m/km. The pressure at point 2 will be:

P_2= 20.36 +1,852m – 1,842m - (20m/km*0.2km) = 26.36m

REACH E

The pipe *e* must transport 4 l/s in total over a distance of 400m, with a residual head of 1.5 to 3 bar (15-30m). We need to find a pipe with a head loss between those values for 4 l/s:

J_{min}= (26.36m + 1,842m – 1,822m -30m)/0.4 km = 40.9m/km
J_{max}= (26.36m + 1,842m – 1,822m -15m)/0.4 km = 78.4m/km

Looking at the tables for 63mm pipe:

45.00	3.752	1.56
60.00	4.396	1.82

For 4 l/s the J value will be between 45m and 60m, and between 40.9m and 78.4m.

To calculate the residual pressure we interpolate to find the answer:

$$\frac{J_x - 45}{60 - 45} = \frac{4 - 3.752}{4.396 - 3.752}$$ J_x = 50.78m/km

This value is in between the previous ones. The residual pressure will be:

P_{F2}= 26.36 +1,842m – 1,822m - (50.78m/km*0.4km) = 26.04m

The mainline will therefore be:

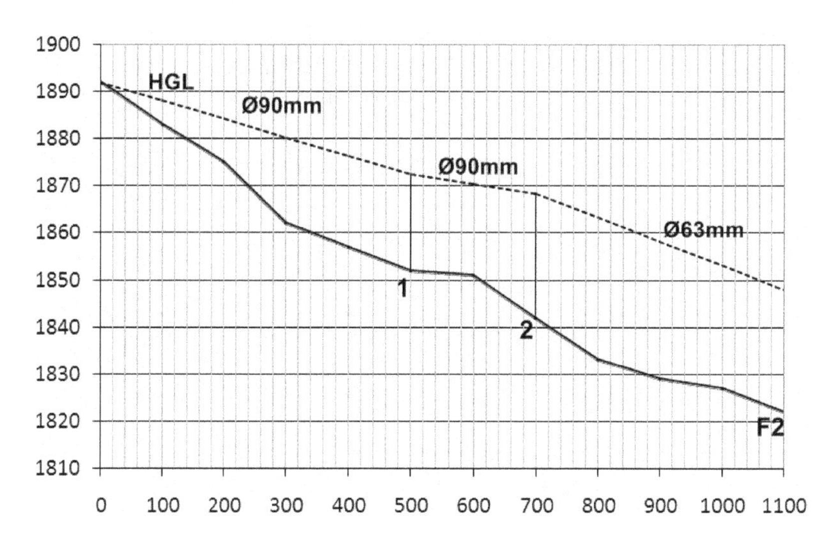

REACH B

Pipe *b* must transport 3 l/s over 500m, to an elevation of 1,846m, beginning at 1,852m. The pressure at *1* has been calculated to be 20.36m.

J_{max}= (20.36m + 1,852m – 1,846m -15m)/0.5 km = 22.72m/km or less.

Looking at the tables for 90mm pipe, we get 5.5m/km. Check the maximum pressure is not exceeded:

P_{F1}= 20.36 +1,852m – 1,846m - (5.5m/km*0.5km) = 23.61m

In case this is excessive, a combination of pipe sizes would have been installed.

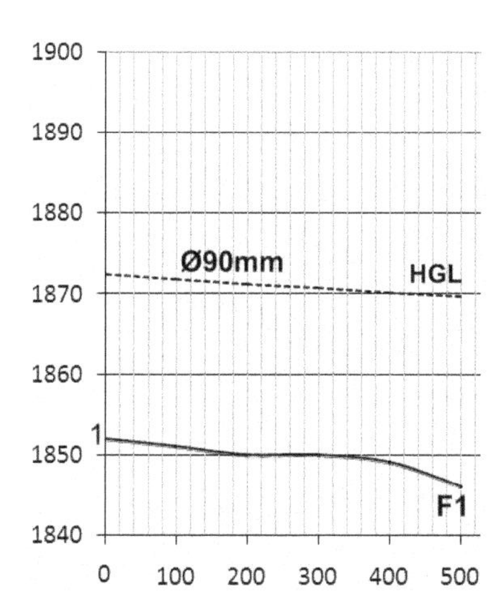

REACH D

The pipe *d* must transport 2 l/s over a distance of 200m, to an elevation of 1850m, departing from 1842m. The pressure at *2* has been calculated as 26.36m.

$$J_{max} = (26.36m + 1{,}842m - 1{,}850m - 15m)/0.2\ km = 16.8m/km \text{ or less.}$$

Looking at the tables for 93 mm pipe, we get 15m/km. Check the maximum pressure is not exceeded:

$$P_{F3} = 26.36 + 1{,}842m - 1{,}850m - (15m/km*0.2km) = 15.36m$$

The partial graph looks like this:

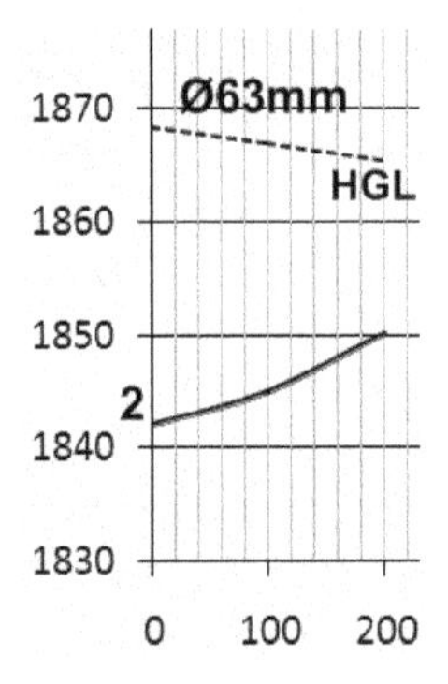

With all the branch lines in the same graph:

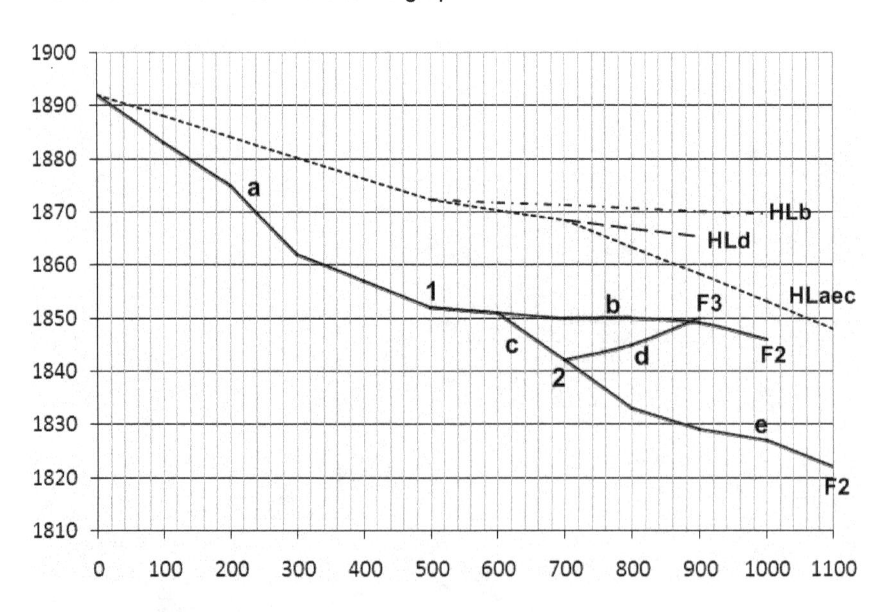

Note that when there´s no demand, for example, at night, the pressure at point F2 is 1,892m-1,822m = 70m. This much pressure at a tap outlet is dangerous and is basically useless for the final users. In the next chapter you´ll see how to resolve this.

2. 9 BURNING OFF PRESSURE

Excessive pressure is as frustrating as a having a tiny flow out of a tap. High pressure produces a mixture of water and air which can be as unruly as a shaken can of soft drink, leaving only a small amount of water once the bubbles have fizzed away. Whoever opens the tap usually gets an unwanted shower and the surrounding area becomes an unsanitary and unsightly mud bath.

Aside from the user points, **it´s best for water systems to work with as little pressure as possible**. More pressure means more problems, burst pipes, higher flow through leaks, more water wasted at point of use and frustrated users. To reduce excess pressure there are 2 solutions: pressure reducing valves, or break-pressure tanks.

A **pressure reducing valve (PRV)** is a slightly more sophisticated solution, which isn´t always that simple or sufficiently robust in the context of cooperation projects. The valve is adjusted at a specific setting. For any system, you take the elevation at which it´s installed together with the value it´s set at. From there, you continue making the calculations as you have done up to now.

A **break pressure tank (BPT)** returns the water to atmospheric pressure by discharging into a tank. This completely depressurises the pipeline, in the same way a puncture flattens a tyre, returning the pressure to 0 as it was when the water first entered the pipe.

Look at section 12.3 in the theory book to find out more.

The fundamental difference between the two is that you can install a PRV in several points in a system and adjust the pressure setting you´re looking for. A BPT is fixed and must be located at a specific elevation above the lowest user point, according to the maximum pressure limits you are working with.

Calculate the HDPE pipe you would need to feed a tap with a maximum of 3 bar, assuming you are using a pressure reducing valve.

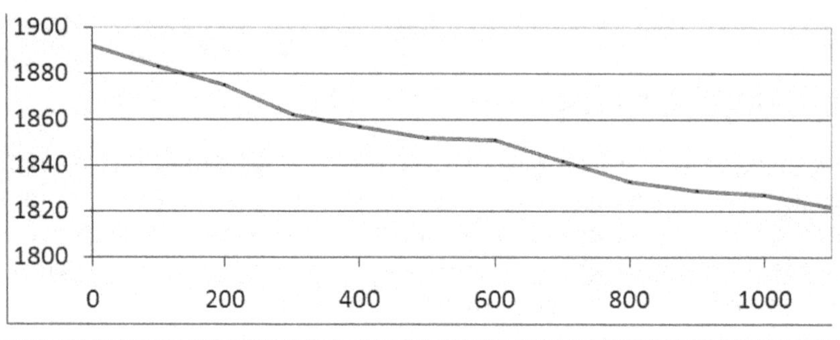

Elevation	1892	1883	1875	1862	1857	1852	1851	1842	1833	1829	1827	1822
Distance	0	100	100	100	100	100	100	100	100	100	100	100
Chainage	0	100	200	300	400	500	600	700	800	900	1000	1100

F2 (4 l/s)

(This is the same topographic profile as in the previous exercise, except you´re only dealing with the mainline).

1. Do the calculations as if there was no PRV. The valve does not affect the pipe diameters: it simply reduces the pressure during times of low demand.

2. The valve should be installed at a maximum elevation of 30m above the lowest user to effectively reduce the energy in the water. If it was installed higher, the water pressure would increase again below the valve. Depending on the distribution of the taps you may need to install several. In this example, one is sufficient.

3. The diameters for a flow of 4 l/s with at least 10m of pressure all along the line are:

   J_{max} = (1,892 - 1,851 -10)m / 0.6 km = 51.66 m/km (for the low point)
   $J_{max\ total}$ = (1,892 - 1,822 -10)m / 1.1 km = 54.54 m/km

   A value of 51.66 m/km or less is fine for the whole reach.
   Interpolating:

	Lower value	x	Higher value
Flow	3.752	4	4.396
J	45	**50.78**	60

$J_{63.\ 4\ l/s}$ = 50.78 m/km.

P = 1,892m – 1,822m – 50.78 m/km * 1.1 km = 14.14m

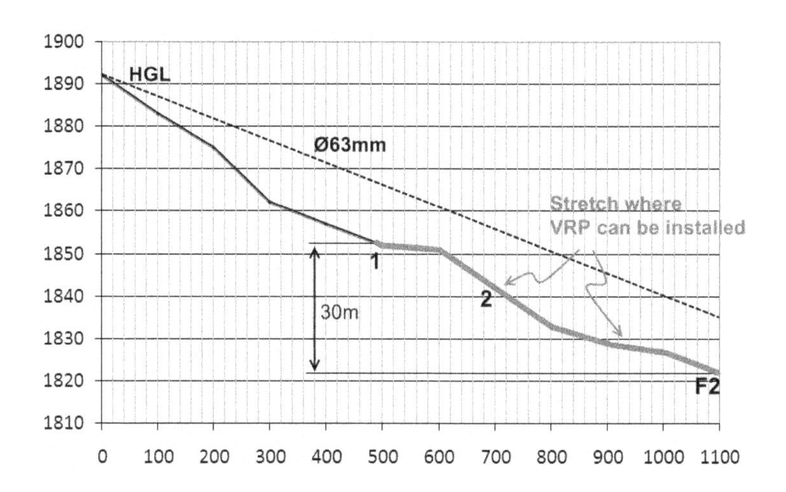

If you were installing a break pressure tank, calculate the same using PVC.

1. The first step is to work out where to place the BPT. If the maximum pressure is 30m, it should be situated 30m above the tap: 1,822m + 30m= 1,852m. This amounts to 500m downstream in the topographic profile.

Take note: this would not be a good place to locate the tank, because the terrain over the next 100m is very flat and the water would need to travel some distance before pressurising again to the minimum of 10m. The best place would be at an elevation of 1,851m, at 600m.

2. The water entering the BPT should arrive with 10m of pressure, to avoid excessive wear of the float valve. So the head loss you´re looking for is:

$$J = (1{,}892m - 1{,}851m - 10m) / 0.6\ km = 51.7\ m/km$$

3. For 63mm pipe at 4 l/s, $J_{63} = 45$ m/km, close enough to avoid combining pipe sizes:

$$P = 1{,}892m - 1{,}851m - 0.6 * 45\ m/km = 14m$$

4. For the pipe leaving the BPT it´s enough to find a diameter that leaves the pressure between 10 and 30m:

$$J_{max} = (1{,}851m - 1{,}822m - 10\ m) / (1.1\ km - 0.6\ km) = 38\ m/km$$

5. In the tables, $J_{90} = 8$ m/km.

$$P = 1{,}851m - 1{,}822m - 0.5\ km * 8\ m/km = 25m.$$

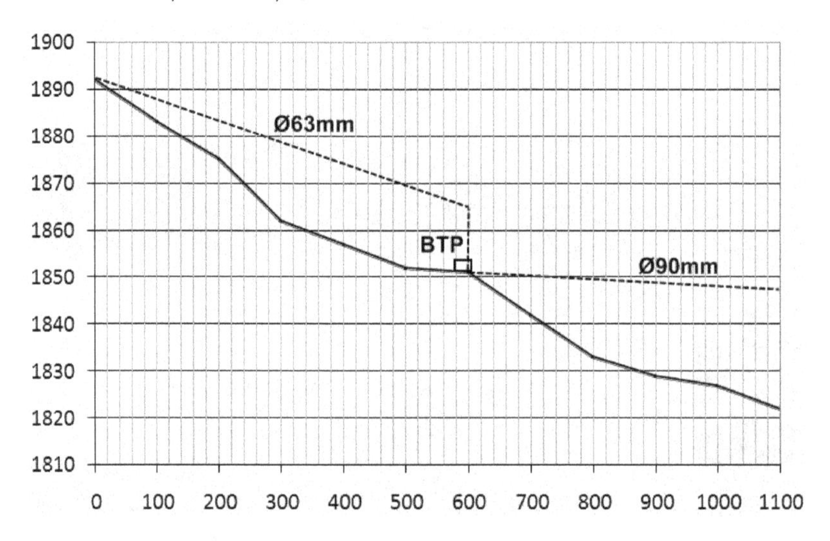

2. 10 PRESSURE ZONES

It´s not always easy to supply an entire area so that all users are within the right pressure limits. People often live on the side of a valley. If the highest users are more than 20m above the lowest, it´s impossible to supply them all within the 10-30m range. If the difference is 30m, for example, when the lowest users have 30m of pressure the highest ones have 0m. If the highest have 10m of pressure, the lowest have 40m.

You´re dealing with a problem which is similar to sleeping with a blanket that´s too short: either you´re feet are freezing or your head gets cold!

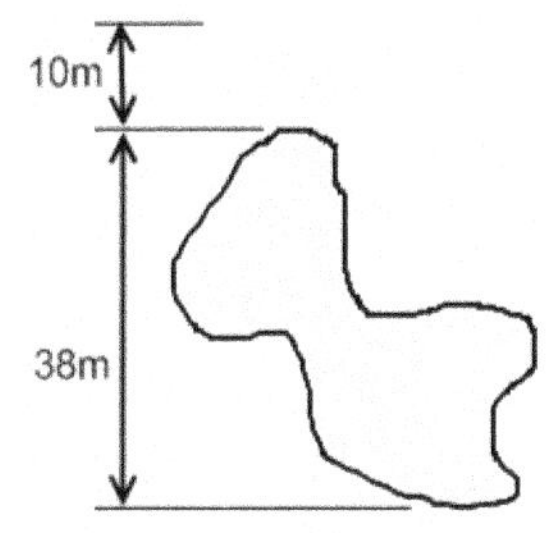

Imagine a system where the source is 10m above the highest user. Between this and the lowest point there´s 38m. With the water at rest, the lowest user has 48m of pressure, almost 2 bar above the maximum. To resolve this problem, the area can be divided in two:

Zone *A*, covers the first 20m, fed directly from the source. Zone *B* covers the following 18m, passing first through a BPT placed 10m above the highest user in B. In each zone the minimum pressure will be 10m and the maximum 30 and 28m respectively.

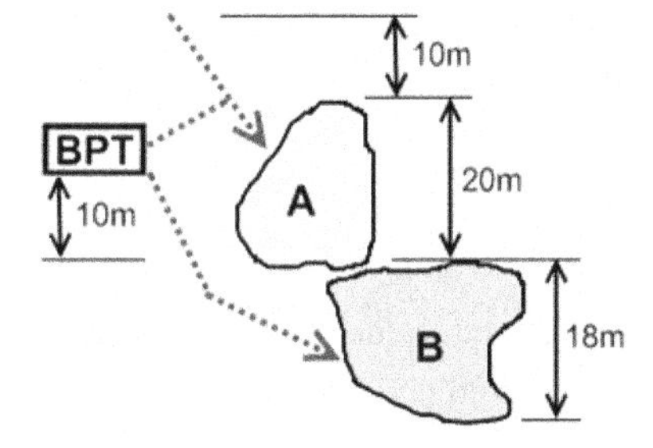

It´s fairly common that in the zone you´re dealing with, the difference in elevation between the high and low points is too great. To avoid making too many zones, a useful rule of thumb is this: make only enough zones to remain within a 35m pressure range.

Creating separate zones can require quite a lot more pipe. In the case that follows, you can´t break the pressure on one hillside as you won´t then be able to get up the one on the opposite side. What you need to do is install piping that feeds the BPT, together with a separate bypass line which maintains the pressure and allows you to reach the opposite hillside:

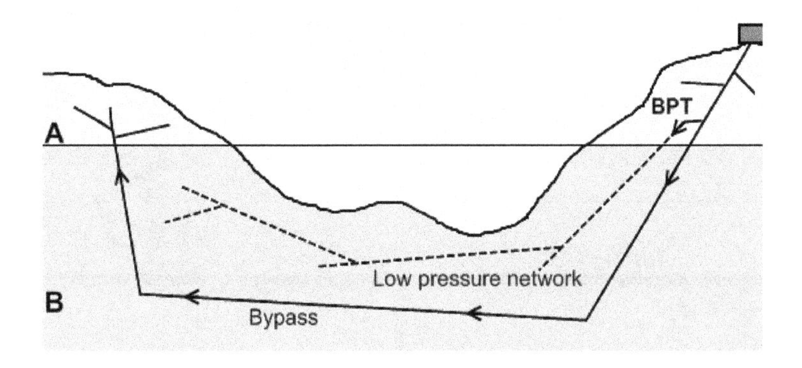

Size the HDPE pipe for this system:

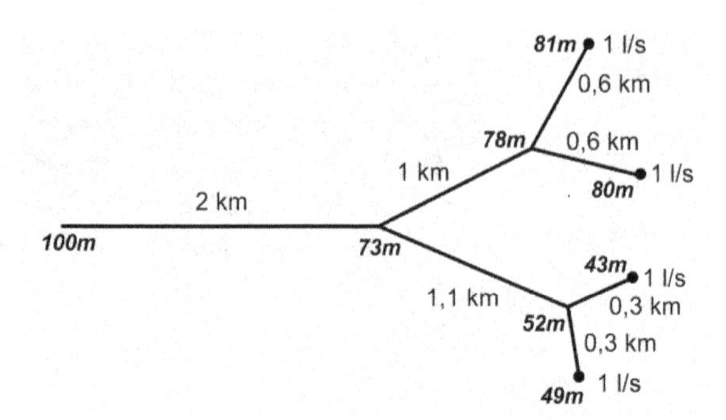

1. The difference between the highest user and the lowest is 81m – 43m = 38m. You´ll have to make two pressure zones, one high and one low.

2. Next, work out where you´ll put the BPT to supply the lower zone. You need to find a site which:

    - allows the water to arrive with sufficient pressure
    - without affecting the piping in the lower zone
    - which doesn´t produce excessive pressure (no more than 30 for the lowest consumer),
    - which gives you a margin for the head loss you´ll have in the pipeline (as close as possible to 30m above the lowest consumer)

In this system, the branch line just after the 73m point fits all the requirements. Below you´ll see where to place the BPT. We´ll also number the pipes:

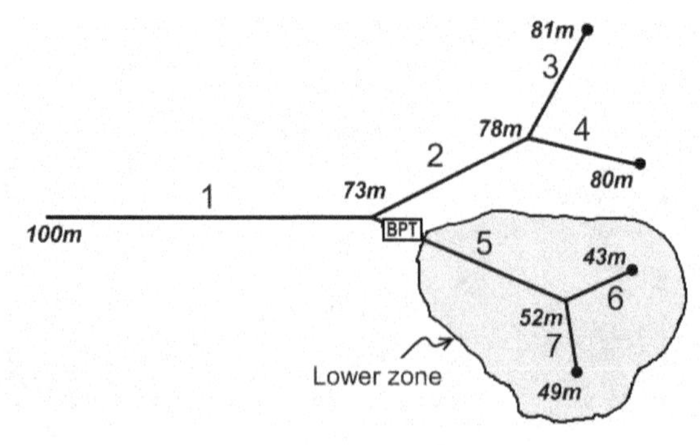

3. No reach will be PN 16, seeing as 100m – 43m = 57m < 80m

Calculating the lower zone.

4. Pipe 5. From the BPT with 73m of pressure, we´re looking for a low-friction pipe:

$J_{90,\ 2\ l/s}$ = 2.5 m/km E = 73m – 1.1 km * 2.5 m/km = 70.25m

P = 70.25m – 52m = 18.25m Ok. → Pipe 5 is 90mm

The efficiency of a pipe relates to its head loss:
→ If it loses less than 5m/km, it´s efficient.
→ If it loses less than 1m/km, it´s super-efficient (and expensive!).
→ Losses of more than 5m/km indicate the need to burn off pressure, or bad design (strangulation).

In the next point, we´ve gone straight for a pipe with a head loss of less than 5m/km, without calculating J_{max} as before. This will save you time on the calculations, once you´re more familiar with the pipes.

5. Pipe 6:

$J_{63,\ 1\ l/s}$ = 4.25 m/km E = 70.25m – 0.3 km * 4.25 m/km = 69m

P = 69m – 43m = 26m Ok.

→ Pipe 6 is 63mm.

6. Pipe 7:

$J_{63,\ 1\ l/s}$ = 4.25 m/km E= 70.25m – 0.3 km * 4.25 m/km = 69m

P = 69m – 49m = 20m Ok.

→ Pipe 7is 63 mm.

Calculating the higher zone.

7. Pipe 1. This needs to be really efficient, as it starts with 73m and then needs to get up to 81m.

$J_{110,\ 4\ l/s}$ = 3.5 m/km (This pipe would leave the pressure only slightly above 81m). Any unauthorized connection or increase in flow could cut off the supply. If the project is very tight on funds and you´ve got no choice in the matter, 110mm is the answer. Whenever you can though, use a pipe which is one size up in these kinds of situations:

$J_{160,\ 4\ l/s}$ = 0.6 m/km E= 100m – 2km * 0.6 m/km = 98.8m

P = 98.8m – 73m = 25.8m

→ Pipe 1 is 160 mm

Remember, there may be intermediate pipe sizes available in the region you´re in, which don´t appear in the tables used here. For example, there may be 140mm pipe.

You can download the complete tables here: www.arnalich.com/dwnl/headloss.zip

8. Pipe 2. This is still the critical section. You need a more efficient pipe.

   $J_{90,\ 2\ l/s}$ = 2.5 m/km E = 98.8m – 1 km * 2.5 m/km = 96.3m
   P = 96.3m – 78m = 18.3m Ok. → Pipe 2 is 90mm.

9. Pipe 3. Again, we need a more efficient pipe here:

   $J_{90,\ 1\ l/s}$ = 0.8 m/km HGL= 96.3m – 1 km * 0.6 m/km = 95.7m

   P = 95.7m – 81m = 14.7m Ok. → Pipe 3 is 90mm.

10. Pipe 4. For reasons of symmetry and similar elevations you´ve probably already got an idea which it is:

    $J_{90,\ 1\ l/s}$ = 0.8 m/km HGL= 96.3m – 1 km * 0.6 m/km = 95.7m

    P = 95.7m – 80m = 15.7m Ok. → Pipe 4 is 90mm.

The result is this:

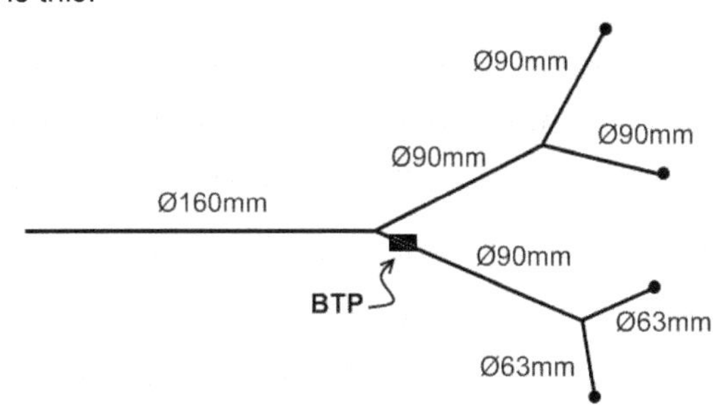

There´s never one right answer for a system, as you will have gathered working through the exercises. There are other criteria which will mean you choose one design over another.

2. 11 MULTIPLE SOURCES

Sometimes you need to use several different sources. Unless you´re incredibly lucky, they are rarely at the same elevation. In these situations, **both sources need to have the same residual pressure when they meet.** When there´s no demand, the higher source will discharge into the lower one. To avoid this, you can install a no-return valve in the pipe line leading to the lower source.

(Intentionally blank)

On one side of a valley a source has been tapped giving 4 l/s at an elevation of 60m (north source). At 52m on the opposite side, a second source provides 2 l/s (south source). Both flows are to be joined at 37m of elevation with HDPE pipe. With the following topographic survey at hand, which pipes should be installed?

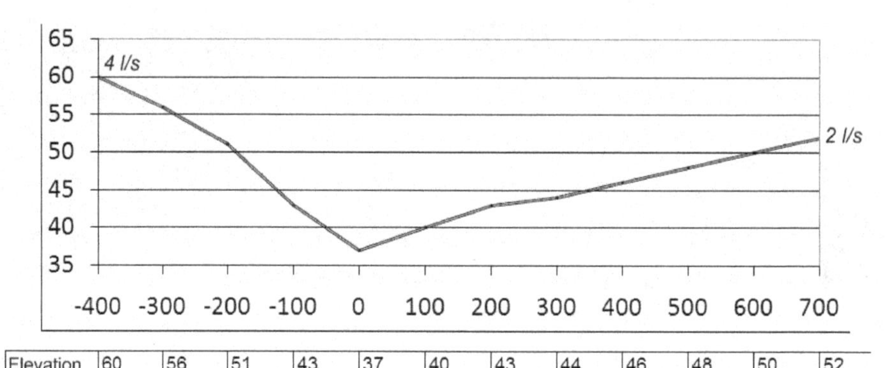

Elevation	60	56	51	43	37	40	43	44	46	48	50	52
Distance	0	100	100	100	100	100	100	100	100	100	100	100
Chainage	-400	-300	-200	-100	0	100	200	300	400	500	600	700
	North S.				**Junction**							**South S.**

We want the pressure at the junction to be greater than 10m, although the topography won´t let us pressurise the system much more than that.

1. The pipe with less room to manoeuvre comes from the south source. Let´s start with this. The maximum head loss is:

   $$J_{max} = (52m - 10m - 37m) / 0.7km = 7.14 \text{ m/km}$$

   For a flow of 2 l/s, the pipe which leaves us enough pressure is 90mm. The pressure at the junction is:

   $$P = 52m - 37m - (2.5 \text{ m/km} * 0.7km) = 13.25m$$

2. The north source must arrive at the junction with 13.25m too. The required head loss is:

   $$J_{north\ source} = (60m - 13.25m - 37m)/0.4km = 24.37 \text{ m/km}$$

   To reach the junction with precisely the required pressure means combining pipe sizes, namely 63mm and 90mm. We can interpolate for 63mm:

$$\frac{J_x - 45}{60 - 45} = \frac{4 - 3{,}752}{4{,}396 - 3{,}752}$$ → J_{63} = 50.78 m/km and J_{90} = 9 m/km.

x * 50.78 m/km + ((0.4km –x) * 9m/km) = 60m -37m – 13.25m

50.78x + 3.6 – 9x = 9.75 → 41.78x = 6.15 → x = 0.147 km of 63mm.

d-x = 0.4 km – 0.147 km = 0.253 km of 90mm.

Note that the order you place the pipes in is very important. If you put the smaller pipe first, the HGL will go underground:

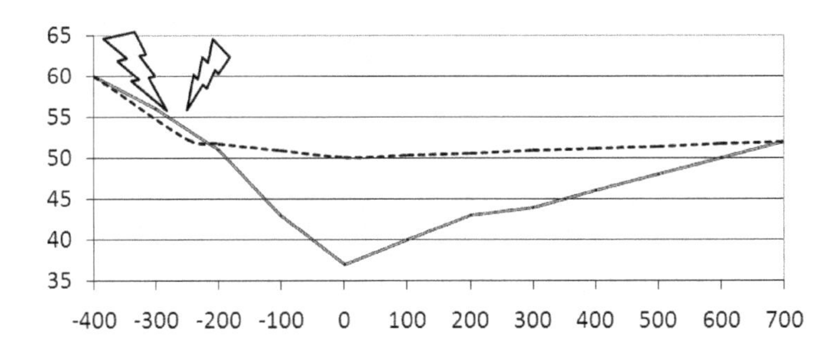

The graph showing the topography and the HGL looks like this:

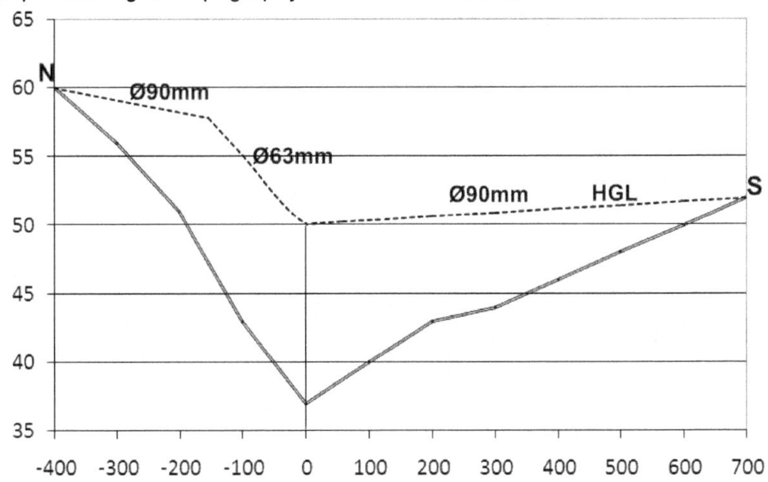

The HGL´s of each pipeline meet at the junction with the same pressure: 13.25m.

2. 13 LOOPED SYSTEMS

Bad news!

Even the simplest looped system cannot be calculated the way we´ve been doing up to now. This is because the water can follow various paths to get to the same point.

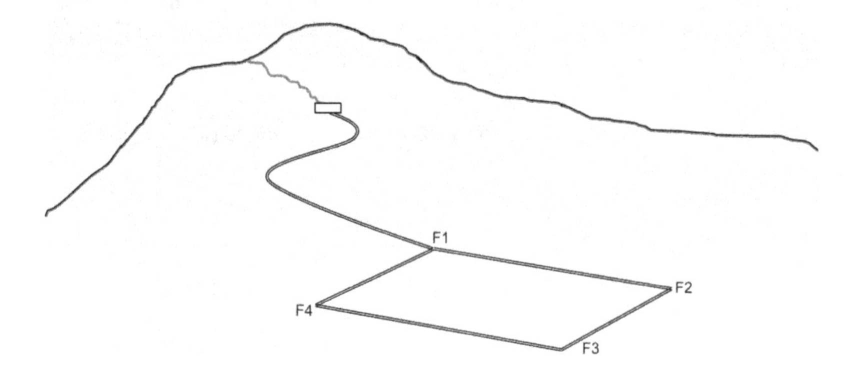

To calculate by hand you can use the **Hardy-Cross method.** It´s more time consuming than complicated for simple systems, and can be as exhausting as having to sit through your brother-in-laws holiday snaps.

I would recommend that instead of losing time learning this method, learn how to do it with a computer from the start. These two books can show you how:

- Arnalich, S. (2007). *Epanet in Aid: How to calculate a water network by computer.* www.arnalich.com/en/libros.html

- Arnalich, S. (2007). *Epanet in Aid: 44 Progressive exercises to calculate water networks by computer.* www.arnalich.com/en/libros.html

If you want a quick tutorial to get over your fear and apprehensions about using a computer for the calculations, chapter 6 of the theory book takes you through an example.

Now that you´ve been warned, this is how to do the calculations by hand, in case you don´t have access to a computer.

THE HARDY CROSS METHOD FOR THE BRAVE

You make successive approximations until 2 conditions have been met:

a. In each node **the mass is conserved**. That means the water coming in is the same as the water going out, and results in all the flows in a node equalling 0.

b. In each loop **energy is conserved,** that means the sum of all head loss in a loop is 0.

The difference between this and what you´ve seen up to now is that this method doesn´t select the pipe sizes for you. It only tells you what happens with the pipes you´ve chosen, i.e. it´s a matter of trial and error. That´s why it becomes a mammoth task, because you´re not only doing the iterations, but you have to keep going over it again and again until you find the optimum pipe sizes. One small change somewhere in the loop and you have to calculate everything over again.

Size the PVC pipes for this system, in which each one is 1km long, and where the nodes all have the same elevation with an input pressure of 3 bar.

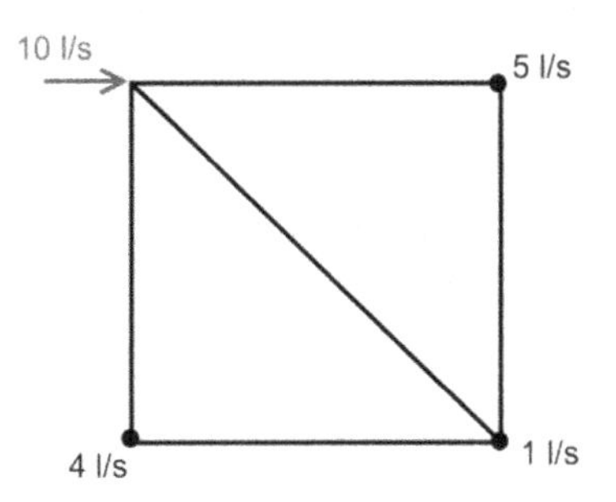

Note: the diagonal is also 1km long. It is represented in the form of a square to make reading easier, although in reality it would be a rhombus.

1. Number each reach and assign a direction of flow to establish if the flow is entering (+) or leaving (-) the node. The loops are also numbered such that each pipe is in at least one loop and in which a direction of flow was be assigned.

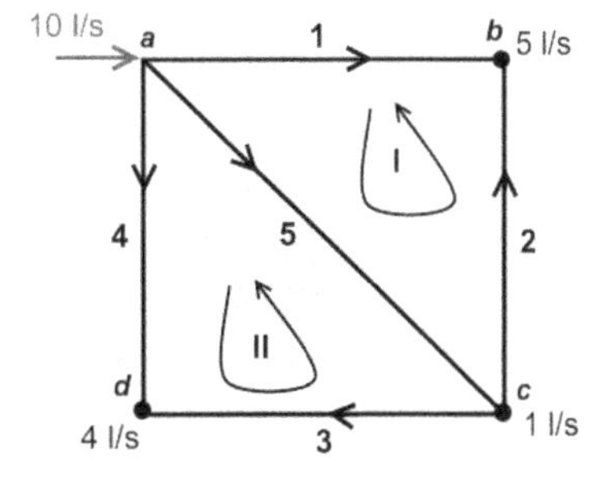

2. Assign a flow to each pipe such that the law of the conservation of mass is respected. But try and fine tune how you distribute the flow, as it´ll save you a lot of calculations. For example:

Pipe 1: - 4 l/s
Pipe 4: - 3 l/s
Pipe 5: - 3 l/s .

In the node *a*, 10 l/s enters and 4 + 3 + 3 leaves. Water entering is positive and water coming out is negative. The balancing equation comes out as:
10 l/s - (4+3+3) l/s = 0 Ok.

Pipe 1: 4 l/s (already assigned)
Pipe 2: 1 l/s in node *b*, (4 +1) l/s -5 l/s = 0 Ok.

Pipe 2: - 1 l/s (already assigned)
Pipe 5: 3 l/s (already assigned)
Pipe 3: 1 l/s in node c, 3 l/s - (1+1+1) l/s = 0 Ok.

Pipe 4: 3 l/s (already assigned)
Pipe 3: 1 l/s in node *d*, (3+1) l/s - (1+1+1) l/s = 0 Ok.

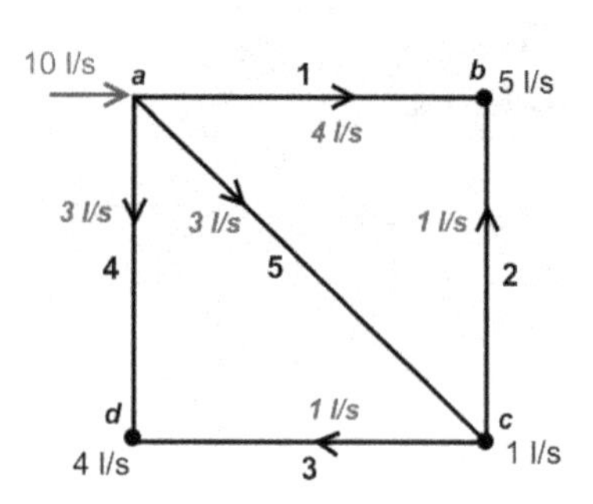

In the first attempt, the flows are:

Check the sum of each node is 0, and that the law of mass is respected.

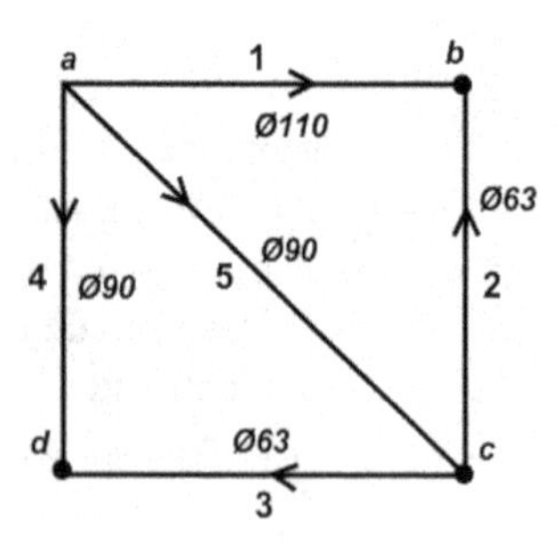

3. With these tentative flows, try a pipe with a diameter that has sufficiently low head loss so as to reach the node in question with an acceptable pressure.
   For example, if we start from 30m down to an elevation of 0m and the longest route in a loop is 2km, we can tentatively establish a head loss of 5 m/km. The pressure we´d be left with is:
   30m – 0m -2 km * 5 m/km = 20m.

4. We can use the tables to find the pipes which have a head loss of around 5m/km for flows of 1, 3 and 4 l/s:

$$J_{63,\ 1\ l/s} = 3.75\ m/km$$

$J_{90,\ 3\ l/s}$ = 4.75 m/km
$J_{110,\ 4\ l/s}$ = 2.75 m/km

5. Let´s see how the loops perform. For loop 1, beginning at *a,* we add the head losses as we move in the direction of the arrows and subtract those which go the opposite way:

1km * 4.75 m/km + 1km * 3.75 m/km – 1 km * 2.75 m/km = 5.75 m/km

5.75 m/km is a little different to 0, which is what the law of loops demands. This means the chosen flows are wrong and they need to be adjusted in the second round.

For loop 2, beginning with *a,* we add the head losses as we move in the direction of the arrow and subtract the ones in the opposite direction. I´ve removed the units here to make it simpler:

1 * 4.75 - 1 * 3.75 + – 1 * 4.75 = -3.75 m/km also different to 0.

6. We need to calculate the flow variations until we get an answer sufficiently close to 0. To find the new flow values and to get their quicker, the old flows are modified in:

$\Delta Q = -\sum h_i / (1.852 * \sum|h_i/Q|)$ Where,

$\sum h_i$, is the sum of the friction produced in each pipe of the loop, i.e. the value you calculated in point 5.

Q = flow

($\sum|h_i/Q|$ is the sum of all the values h_i/Q taken as positive)

For the first loop:

$\sum|h_i/Q|$ = 1* 4.75/ 3 + 1* 3.75/1 + 1* 2.5/4 = 6.02

ΔQ_I= - 5.75 / (1.852 * 6.02) = - 0.52 l/s

For the second:

$\sum|h_i/Q|$ = 1 * 4.75/3 + 1 * 3.75/1 + 1 * 4.75/3 = 6.92

ΔQI_I= - (- 3.75 / (1.852 * 6.92)) = + 0.29 l/s

When a pipe belongs to two loops, the correction is made for both, i.e. for pipe 5, the correction will be -0.52 + 0.29 = - 0.23 l/s.

If the flow goes in the direction of the arrow it´s considered positive, otherwise negative. The variation ΔQ is always added up.

7. The flows are corrected:

Pipe 1: - 4 – 0.52 = - 4.52 l/s
Pipe 2: 1 – 0.52 = 0.48 l/s
Pipe 3: - 1 + 0.29 = - 0.71 l/s
Pipe 4: 3 + 0.29 = 3.29 l/s
Pipe 5: 3 – 0.52 + 0.29 = 2.73 l/s

This marks the end of the first iteration; now with these flows you need to repeat the cycle you began in point 2.

Just let me run off here and give you the values calculated by a computer, in case you really want to carry on with the calculations by hand. Ignore the arrows, the computer hasn´t necessarily chosen the directions the way you have. You can consider the flow values to be acceptable when the variation is less than 5%, i.e. $\Delta Q < 0{,}05 * Q$.

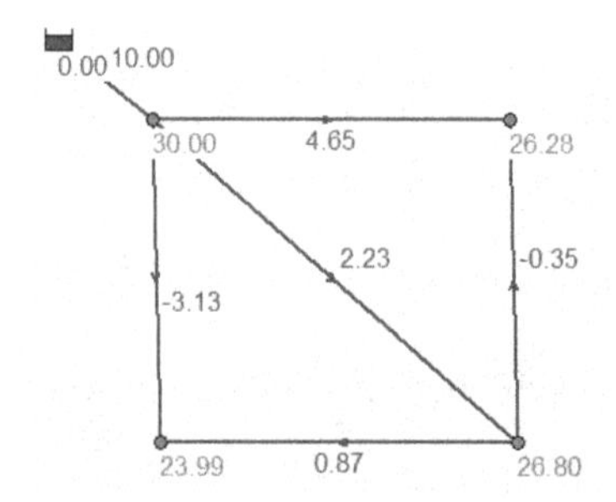

Once you´re done, to calculate the pressure values, just take any route towards a node and calculate the head loss along the whole route, as you have done in the previous exercises. Any route has the same head loss, as the system is in equilibrium.

Notice how the pressures are 23.99m, 26.28m y 26.8m, which are acceptable. All these calculations go to show that the initial chosen diameters work. This doesn´t mean they are necessarily the optimum choice.

2. 12 SPREADSHEETS

All the calculations we´ve done so far in this book are made much easier with the use of a spreadsheet like Excel, which also allows you to draw graphs. Since it´s the most commonly used, the explanations here will be with Excel 2007 for Windows, although there are plenty of free open-source alternatives like OpenOffice Calc:

 http://www.openoffice.org/

If you´ve got a computer to hand, and you think that in the future you´re likely to design more systems, your best bet is to learn how to use a modelling program like Epanet.

Here we assume a basic working knowledge of Excel. If you´re starting from scratch, download this short tutorial:
www.arnalich.com/dwnl/Exceltuteng.pdf

Once you´ve entered the data into the spreadsheet, you can create graphs similar to the one below, where the pressure, HGL and topography are displayed.

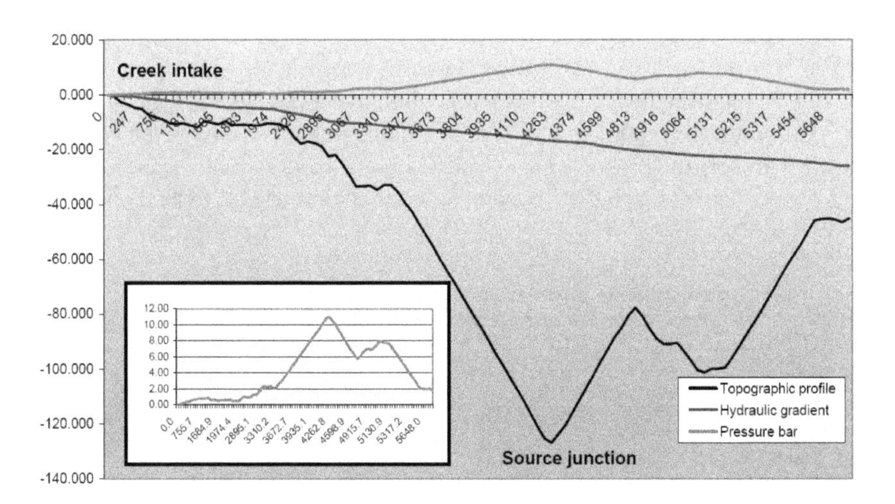

Size the pipes for the mainline supplying the refugee camp Mtabila II using HDPE with a flow of 9.8 l/s:

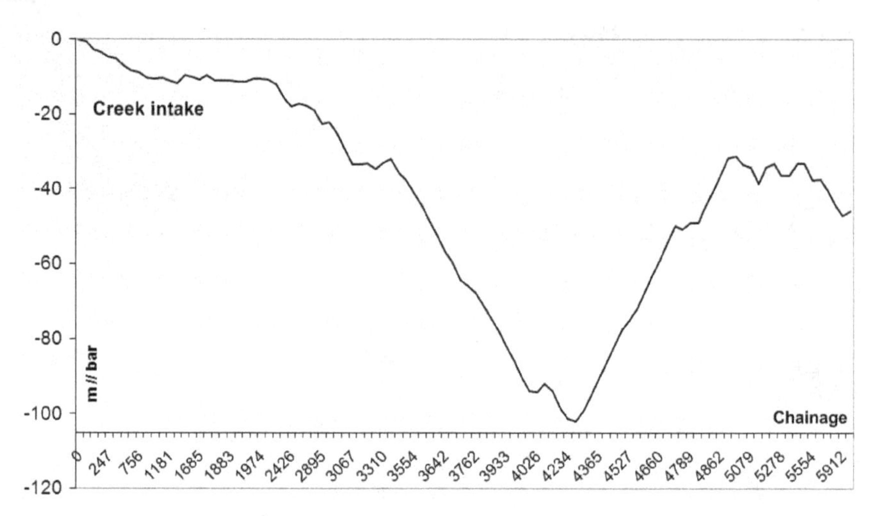

The first step is to prepare an Excel sheet to enter in the data:

1. Open Excel of the spreadsheet.

2. On the page heading, enter a title which makes sense and include information regarding the date, who prepared the document, the version number, and probably some kind of logo of the organisation you´re working for. The idea is that the document be clearly identifiable to third persons.

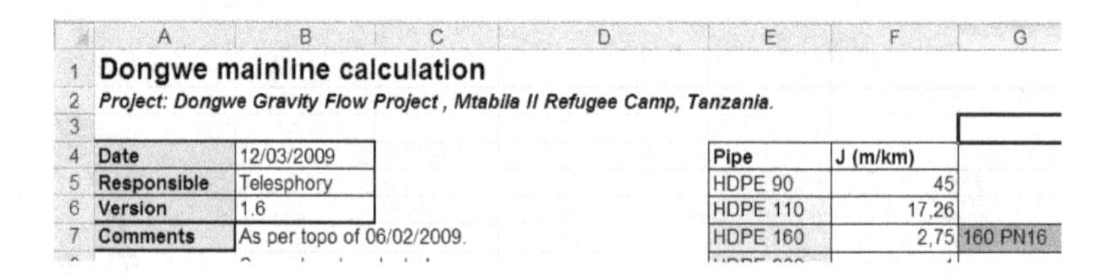

Dongwe mainline calculation

Project: Dongwe Gravity Flow Project , Mtabila II Refugee Camp, Tanzania.

Date	12/03/2009
Responsible	Telesphory
Version	1.6
Comments	As per topo of 06/02/2009.

Pipe	J (m/km)	
HDPE 90	45	
HDPE 110	17,26	
HDPE 160	2,75	160 PN16

3. Save the document **with the same name as the title** you´ve used:

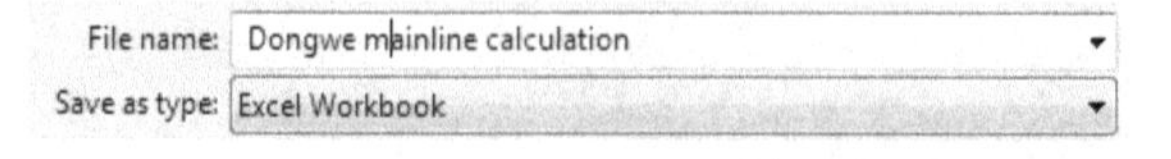

You´re not going to spend the rest of your life working on this same system: there will be a moment when other people inherit the project. There´s nothing more frustrating than trying to fight your way through a million files called “water 1,” “Peter,” “Dongwe test” and so on...In the case of version numbers, it´s vital to be really organised so that no one else starts using a document that´s been only half worked on.

4. On one row, enter the column headings:

- Distance
- Chainage
- Elevation
- Notes (to include any comments)
- J (head loss)
- J. cum (cumulative head loss)

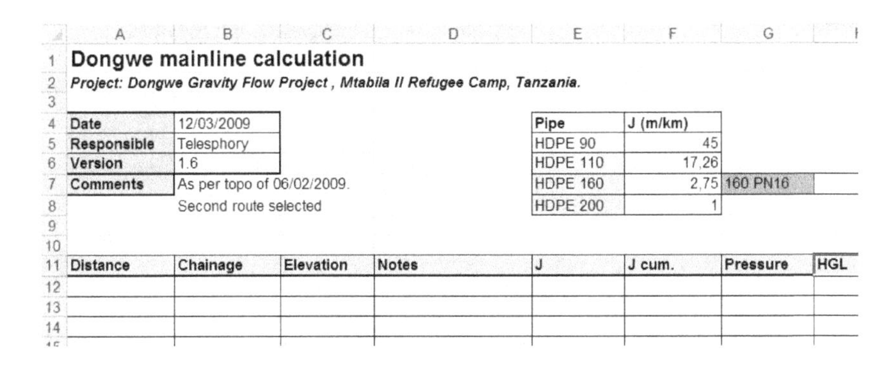

Dongwe mainline calculation

Project: Dongwe Gravity Flow Project , Mtabila II Refugee Camp, Tanzania.

Date	12/03/2009
Responsible	Telesphory
Version	1.6
Comments	As per topo of 06/02/2009.
	Second route selected

Pipe	J (m/km)	
HDPE 90	45	
HDPE 110	17,26	
HDPE 160	2,75	160 PN16
HDPE 200	1	

Distance	Chainage	Elevation	Notes	J	J cum.	Pressure	HGL

5. Enter the Distance and Elevation values from the topographic survey. You don´t need to copy the Chainage as it´ll be calculated automatically.

Distance	Chainage	Elevation	Notes	J	J cum.	Pressure
0		0	Toma en el arroyo			
91,24		-0,596				
76,9		-2,607				
67,2		-3,478				
11,8		-4,672				
139,8		-5,109				
154,2		-7,222				
118,9		-8,241				
95,70		-8,911				

To avoid having to manually enter values or trying to work them out from the graph, download this file in which they´ve already been entered:

www.arnalich.com/dwnl/19topoeng.xls

Once you´ve opened it check the last row:

111	114,22	-44,275
112	0,00	-47,152
113		
114		

The distance is zero, and the difference in elevation is 45.152 - 44.274 = 2.877m. This corresponds to the entrance of the tank, which is generally from above.

The elevations are negative because the highest point has been arbitrarily taken to be at an elevation of 0m.

6. Calculate the chainage. To do this, enter the formula "=B12+A13" into cell B13. Don´t forget the "="! This adds the new Distance value with the chainage each time.

J12 =B12+A13

	A	B	C	D
9				
10				
11	**Distance**	**Chainage**	**Elevation**	**Notes**
12	0		0	Intake
13	91,24	=B12+A13	-0,596	
14	76,9		-2,607	
15	67,2		-3,478	

7. To complete the rest of the cells, write 0 in B12 (chainage at the beginning is 0). Then place the cursor over the bottom right hand corner of the cell and note how it changes from a white cross to black one. This means you can drag and copy the formula down the column into the cells bellow.

1	**Distance**	**Chainage**	**Elevation**
2	0		
3	91,24	91,24	-0,59
4	76,9		-2,60

8. Click on the corner and without letting go, drag the cross down to the last cell. All the cells you´re copying the formula to, will be surrounded by a grey

dotted box. When you let go, the formula will calculate the chainage in each cell. The final value should be 5,911.93m. The pipe line will be almost 6km long.

If you click on the cell B14 you´ll see the formula changes, rising by 1 each time from B12+A13 to B13+A14 and so on, in all the cells below.

POTENCIA =B13+A14

	A	B	C	D
10				
11	Distance	Chainage	Elevation	Notes
12	0		0	Intake
13	91,24	91,24	-0,596	
14	76,9	=B13+A14	-2,607	
15	67,2	235,36	-3,478	
16	11,8	247,11	-4,672	
17	139,8	386,91	-5,109	
18	154,2	541,11	-7,222	
19	118,9	659,97	-8,241	

9. In the tables find the head loss for 9.8 l/s for different diameters. Create an excel table:

$J_{90;\ 9.74\ l/s}$ = 45 m/km
$J_{110;\ 9.8\ l/s}$ = 17.26 m/km (interpolated value)
$J_{160;\ 9.71\ l/s}$ = 2.75 m/km
$J_{200;\ 9.8\ l/s}$ = 1 m/km

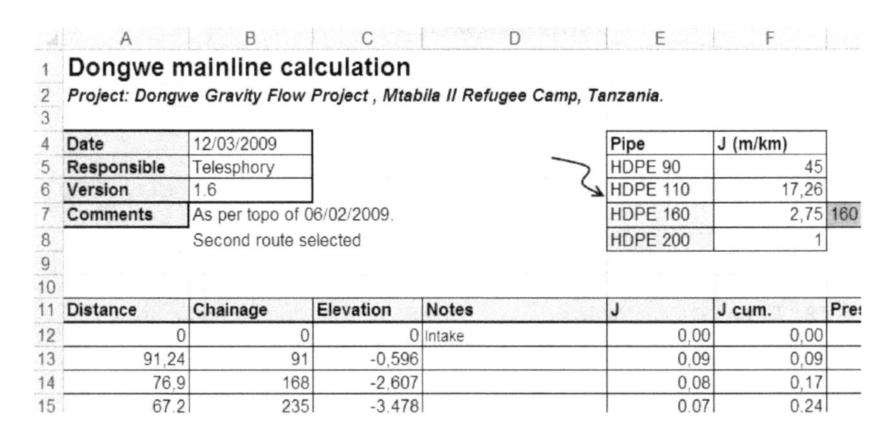

Dongwe mainline calculation

Project: Dongwe Gravity Flow Project , Mtabila II Refugee Camp, Tanzania.

Date	12/03/2009
Responsible	Telesphory
Version	1.6
Comments	As per topo of 06/02/2009. Second route selected

Pipe	J (m/km)	
HDPE 90	45	
HDPE 110	17,26	
HDPE 160	2,75	160
HDPE 200	1	

	A	B	C	D	E	F	
11	Distance	Chainage	Elevation	Notes	J	J cum.	Pre
12	0	0	0	Intake	0,00	0,00	
13	91,24	91	-0,596		0,09	0,09	
14	76,9	168	-2,607		0,08	0,17	
15	67,2	235	-3,478		0,07	0,24	

10. You can find the head loss in each reach, J, by multiplying the Distance by the head loss of the pipe you´ve selected in the table you´ve just made. To do this, enter the following formula into cell E12: "=A12*$F$5/1,000".

	Pipe	J
	HDPE 90	45
	HDPE 110	17,26
2/2009.	HDPE 160	2,75
a.	HDPE 200	1

tes	J	J. Cum	Pressure
	=A12*$F$5/1000		
	4105,8		

The $ symbol ensures that the reference always points to the same cell, F5, even if you copy it down into other cells. Dividing by 1,000 is required as the distances are in meters but the head loss is in kilometres.

11. Copy the formula until you´ve got the values for each reach. You´ll probably find it annoying to have more than two decimal points, so right click anywhere on the screen and select Cell Format. Once you´re there, select the Number tab, then the Number category and write 2 in the decimals box:

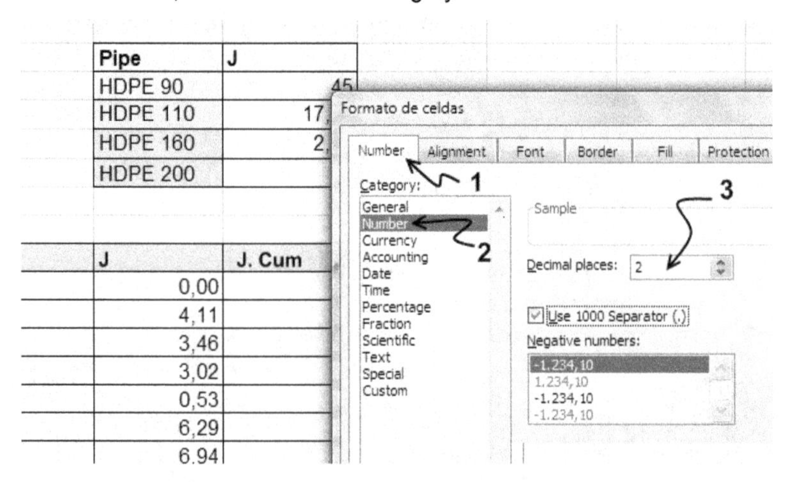

12. To find the cumulative J, do the same as in point 6 and drag down:

	HDPE 200	1	
	J	**J. Cum**	**Pressure**
/o	0,00	0,00	
	4,11	=F12+E13	
	3,46		

13. Now the pressure column. The pressure is the initial elevation, less the elevation at one particular point, less the head loss. Enter this into the formula:

	B	C	D	E	F	G
10						
11	**Chainage**	**Elevation**	**Notes**	**J**	**J cum.**	**Pressure**
12		0	Intake	0,00	0,00	0
13	91,24	-0,596		4,11	4,11	=$C$12-C13-F13
14	168,2	-2,607		3,46	7,57	
15	235,36	-3,478		3,02	10,59	

The initial pressure is 0. $C$12 is the initial elevation, and the $ symbol is needed so it points to the same cell when you drag and copy. C13 is the elevation at that point, and F13 the Cum. J, or cumulative head loss.

14. Drag and copy the formula down and read the pressure values. You´ll notice that they are all negative values. The 90mm pipe you chose is too small:

J	J. Cum	Pressure
0,00	0,00	0
4,11	4,11	-3,51
3,46	7,57	-4,96
3,02	10,59	-7,11
0,53	11,12	-6,45
6,29	17,41	-12,30
6,94	24,35	17,12

To be able to see the results more easily, it´s best to make a graph.

15. The line you´ll pay closest attention to in the graph is the HGL. Seeing as it doesn´t exist yet, create an HGL column next to the pressure column. The value is the initial elevation less the cumulative head loss.

	J cum.	Pressure	HGL
0,00	0,00	0	=$A$12-F12
4,11	4,11	-3,51	
3,46	7,57	-4,96	

Drag and copy the formula to get the remaining values. As the initial elevation is zero, the values are the same as cumulative J but with the sign changed.

16. Select the data in the Elevation and chainage column. Click on the centre of cell with the Elevation title, and without letting go, drag down. You´ll see the cells are selected, and should be shaded:

Distance	Chainage	Elevation	Notes	J	J cum.
0	0	0	Toma en el arroyo	0,00	(
91,24	91	-0,596		4,11	4
76,9	168	-2,607		3,46	7
67,2	235	-3,478		3,02	1(
11,8	247	-4,672		0,53	11
139,8	387	-5,109		6,29	17
154,2	541	-7,222		6,94	24
118,9	660	-8,241		5,35	29
95,70	756	-8,911		4,31	34

17. In the upper menu bar of Excel, go to Insert, then Dispersion, and then the graph type number 3:

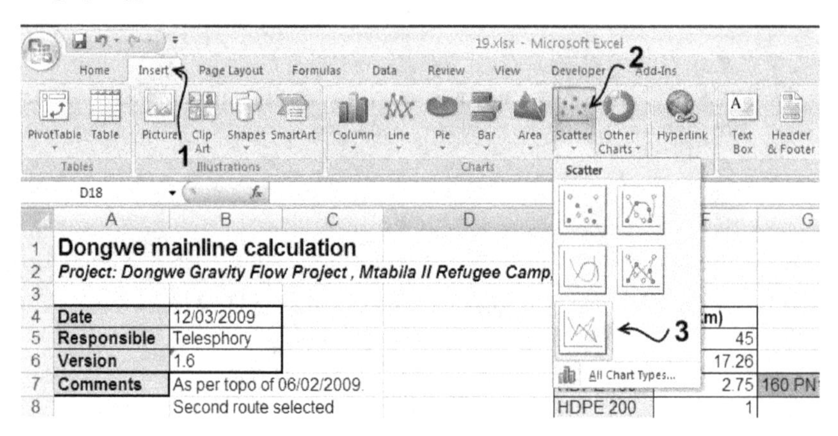

18. The graph should be something like this:

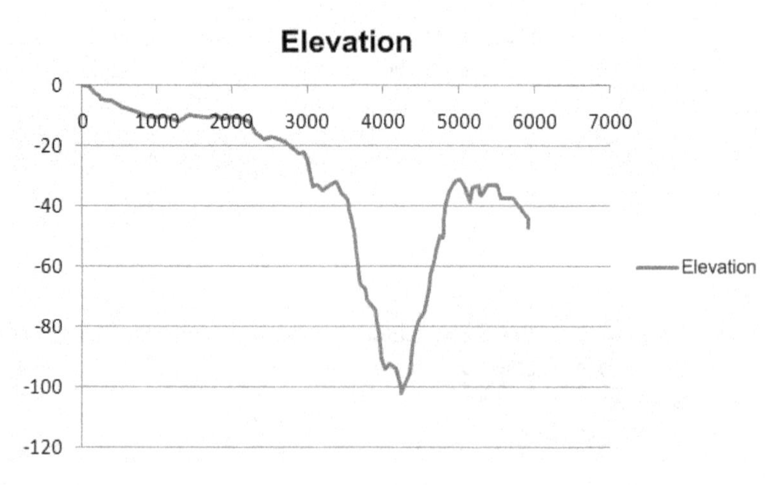

19. The HGL still needs to be plotted. Right click on any blank area on the graph and choose Select data. You´ll see a box like the one below, where you have to select Add:

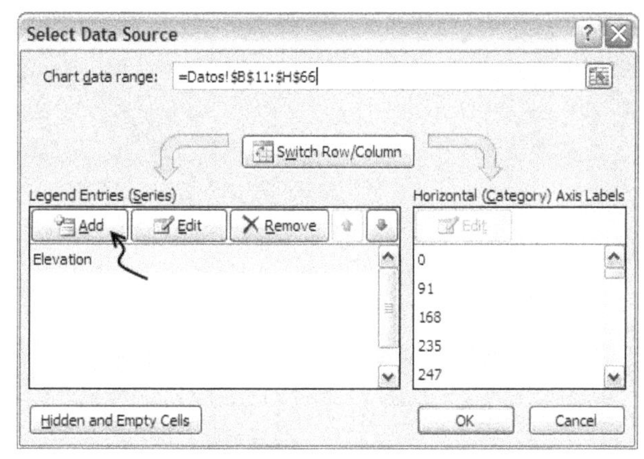

20. Write HGL in the first box, and click on the table symbol in the second:

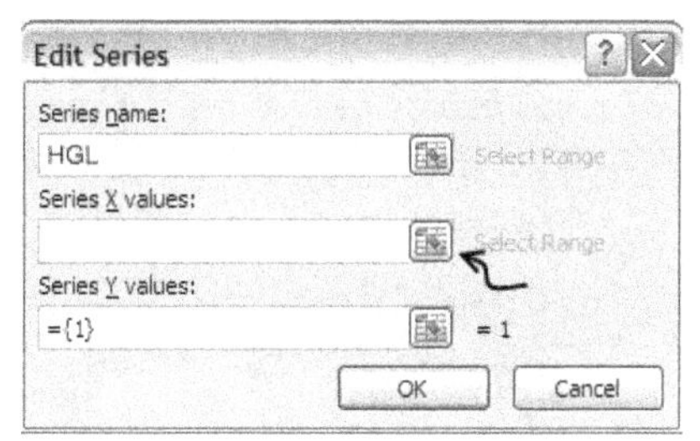

21. Select the cells which have the chainage values and click on the symbol to accept:

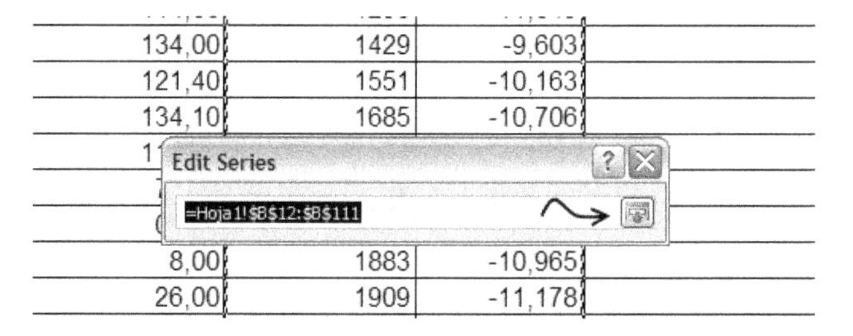

22. Repeat for the Y axis with the HGL values, so that the box looks like this:

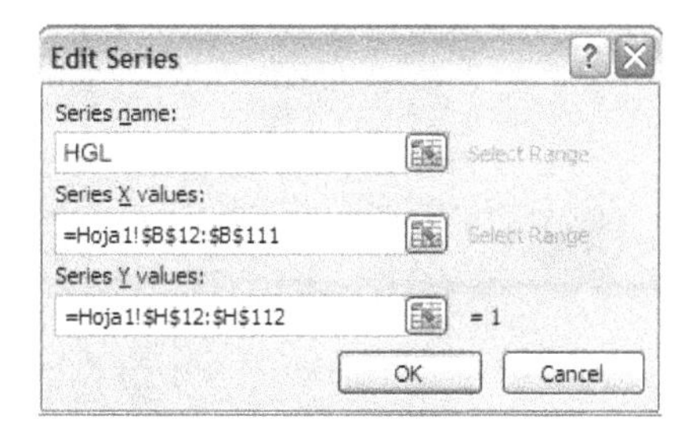

The graph should look like this. You´ll recognise it from previous exercises:

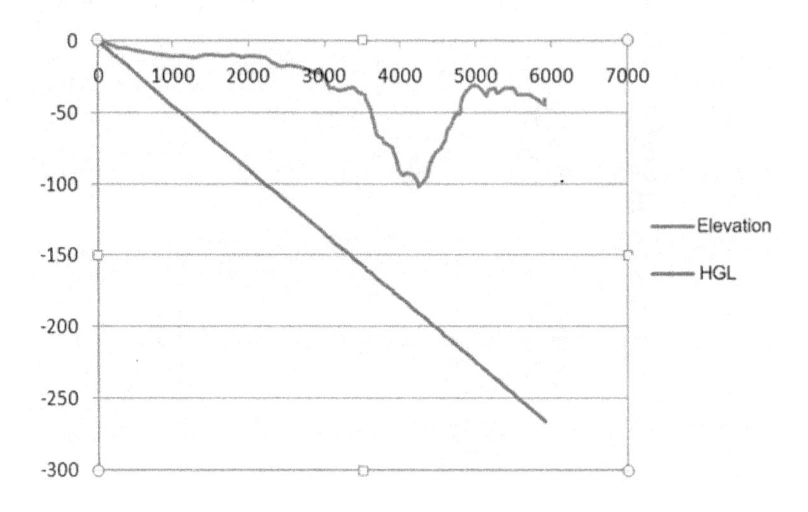

You can make a series of improvements, until you get the result we´re using here. But since this is not an Excel manual, I´ll let you work out how to do it on your own:

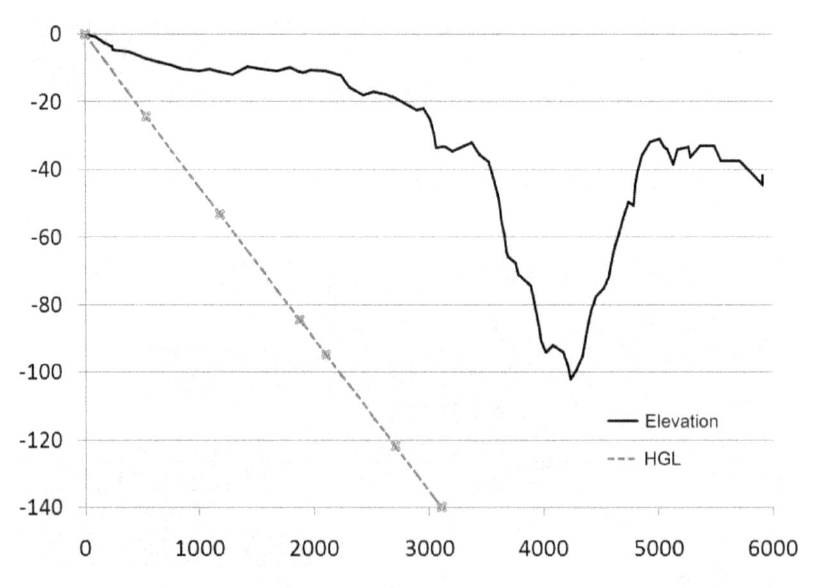

If you got lost at any point, you can download the Excel document, with what we´ve done up to now. It can also be useful as a template:

www.arnalich.com/dwnl/gravitytemplate.xls

Now select the pipes until the HGL is 10m above the terrain at all points, with a pressure of between 10 and 30m.

23. Notice that there´s a critical point 2,241m from the origin. Up until that point we want the HGL to be as flat as possible, so we can maintain 10m of head and pressurise the pipe as fast as possible. Modify the formulas from distance 0 up to 2,241m, so that they point at the 200mm pipe. This means changing $F$6 for $F$8 in the formula, then drag and copy down to the distance of 2,241m (row 38).

Pipe	J
HDPE 90	45
HDPE 110	17,26
HDPE 160	2,75
HDPE 200	1

	J	J. Cum	Pressure	HGL
	=A12*$F$8/1000		0	
	4,11	4,11	-3,51	

24. To check the pressure at the 2,241m point, read off the value in the Pressure column: 9.70m. It´s not quite 10m but it´s close enough. The problem is the topography, which doesn´t allow for any more. Increasing the pipe size will increase costs for very little result. Check out how the graph has changed:

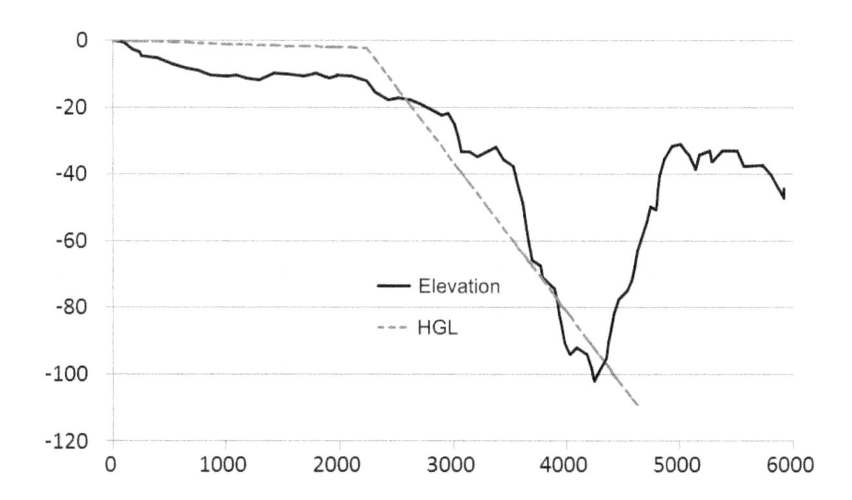

25. For the next reach, notice how the difference in height is very pronounced and is going to require PN16 at the bottom of the valley. This means the next section will leave from 2,241m until the pressure reaches 80m. Starting from zero, the point at which the elevation is -80m is at a distance of 3,912m (-78.011).

26. Choose the pipe for the reach which goes from 2,241m to 3,912m. To do this, multiply the first cell after the 2,241m point by the head loss for the pipe you want to try. In this case, let´s try 160mm (cell $F$7):

,44	2104	-10,738	0,13
,50	2241	-11,939	0,14
,90	2313	-15,58	=A39*$F$7/1000
25	2426	-17,882	5,10

27. Copy and drag the cells to 3,912m and check the result:

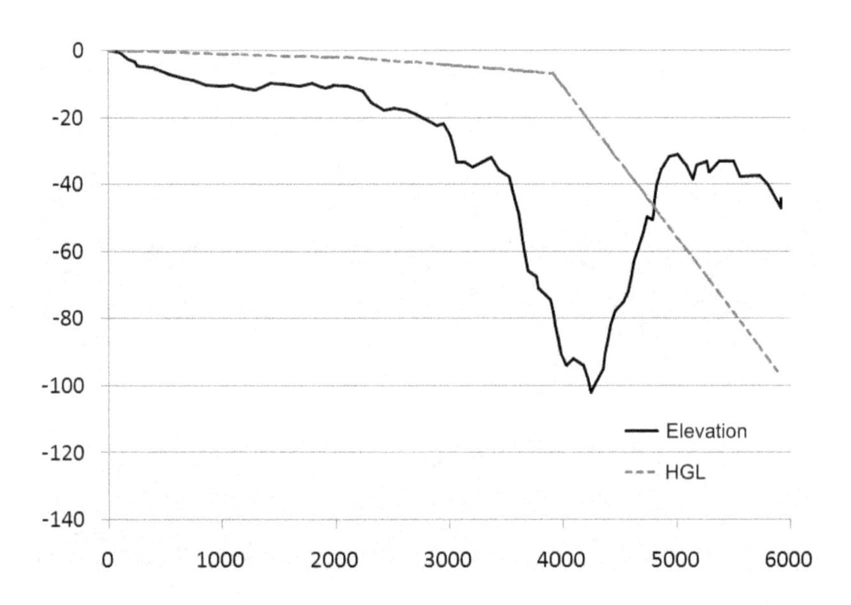

It looks pretty good! If you lengthened the reach you´ve just done, you´d get to the end of the line with sufficient pressure.

28. From 3,912m to 4,459m, where the pressure falls back below the 80m mark (elevation of -77.54m), you´ll need to install PN16 pipe. Seeing as the pipe you used in the previous reach looks good, try the same diameter here but with PN16, and see what happens. To do this, go back to the tables, find the 160mm value for PN16, and incorporate it into your Excel table:

$$J_{160mm,\ PN16,\ y\ 9.782\ l/s} = 4\ m/km$$

Pipe	J		
HDPE 90	45		
HDPE 110	17,26		
HDPE 160	2,75	160 PN16	4
HDPE 200	1		

29. Multiply the first cell after 3,912m by the head loss for the pipe you´ve just introduced (in cell $H$7). Drag and copy the cells up to 4,459m and look at the result:

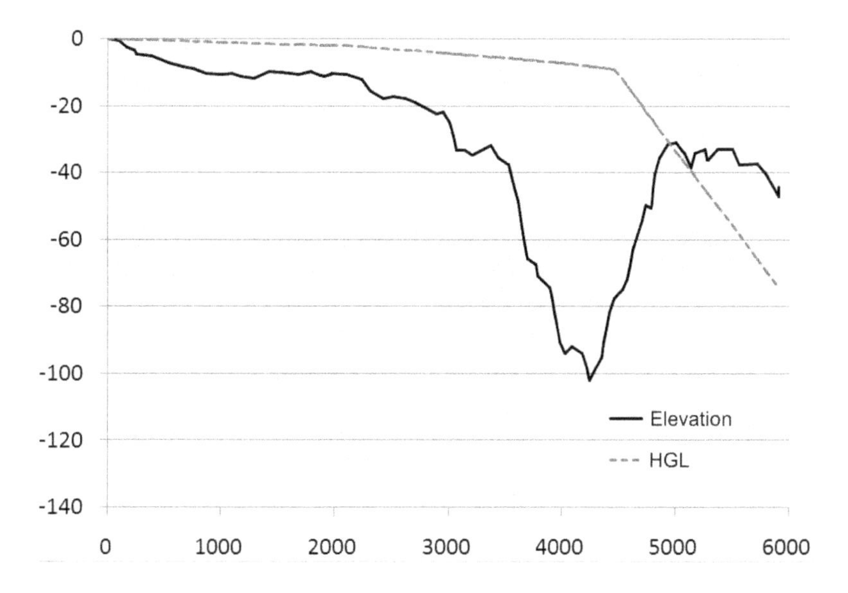

30. Do the final reach with 160mm PN10, and read off the pressure at the entrance to the tank: 31.28m.

31. This is too much pressure. You´ll also save money using smaller pipe. To reduce the pressure, go to the last cell (where the water enters the tank) and make it point to 110mm pipe ($F$6). Then copy upwards and note what happens with the pressure at the end point and in between.

32. You can get to approximately 4,800m. The graph looks like this:

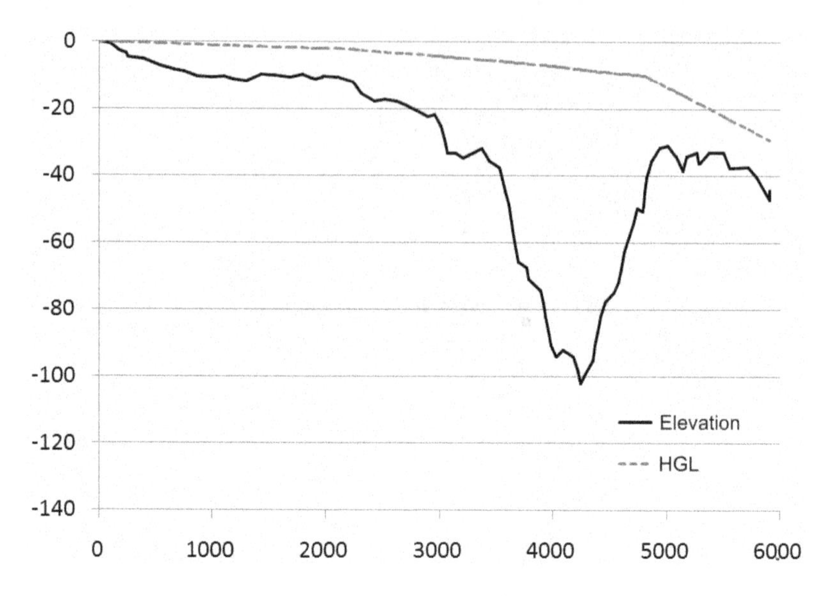

33. Congratulations, the calculation´s done! Now use the notes column to include information such as "Start of 160mm pipe," or "End of 160mm PN16 pipe," and so on. Note that I´ve put a colour scale in the friction loss table for each pipe. You can indicate which reaches use which kind of pipe by shading the background of the table with these colours.

The different reaches end up being like this:

0- 2,241 m	HDPE 200mm PN10
2,241 – 3,912m	HDPE 160mm PN 10
3,912 – 4,459m	HDPE 160mm PN16
4,459 – 4,789m	HDPE 160mm PN10
4,789 – 5,912m	HDPE 110 mm PN10

 You can see the result by clicking on this link (click on the image itself to enlarge): www.arnalich.com/dwnl/19zheng.png

You can also download the completed spreadsheet here:

www.arnalich.com/dwnl/19eng.xls

It may seem kind of laborious to use a spreadsheet if you´ve never done so before. Just remember it has some of the following advantages:

1. You have clear information with all the relevant data.
2. You have a graph which is easy to interpret.
3. You can easily share the information as it´s now in digital format.
4. Mistakes are easier to detect and can be quickly sorted out.

Be careful when you use spreadsheets as they can be deceptive. If you select the wrong data in a formula, or get the cells mixed up, your results can end up being wrong. The good news is they are easy to detect and correct, as you´ve just found out.

2. 14 PRESSURE AS A DESIGN CRITERION

Now that you´re settling in a little, it´s worth making a couple of points which may have got in your way before:

- **The height of buildings.** The pressure at the tap should be between 10 and 30m, which won´t be at street level. This means that if you've got a building that´s 3 stories high, you´ll have to take into account the building height, and provide 10m of pressure on the top floor: 10m + 3* 2.5m = 17.5m above street level. Follow the same reasoning for buildings up to 5 stories high. Anything more than that, and the buildings themselves will have to take care of assigning a pressure group for the users on the highest floors.

- **Pressure for unforeseen circumstances.** Branched systems are very sensitive to mistakes in flow estimations. To build more robust and expandable systems, try aiming for the upper range of the design margin (i.e. 20-30m). The problem with this is that it often requires a large investment, in which case, the 10-20m range is acceptable while not necessarily desirable.

- **Leaks and maintenance.** The flow from a leak increases exponentially with the pressure. So do the chances of a pipe breaking. That means it´s best to keep the pressure as low as it can be for the system to function properly.

- **Head loss in accessories.** There are not only frictional losses in pipes, but in elbows, T´s, valves and other accessories, causing a loss of pressure, known as *minor losses*. They are only relevant for high velocities, and generally don´t have much effect on traditional gravity flow systems.

You can read more about this in section 5.5 of the theory book.

- **The water hammer.** The pressure in a pipe increases if a valve is shut off very suddenly. The pipes can burst or collapse on themselves due to the shock waves. Wherever valves are installed to cut off the flow, you need to calculate the water hammer, to make sure it does not exceed the maximum rated pipe pressure.

You can read more about this in section 7.4 of the theory book.

3. Demand and flow

In all the exercises up until now, you´ve been given a specific flow to work with. In this chapter you´ll learn how to work out and establish a flow, with which to begin designing your system.

3. 1 BASE DEMAND

This is the quantity of water which the popluation will consume for all kinds of different uses: cooking, washing, drinking, work activities and so on...There is no quick fix to working out the demand of the specific population. As a general pointer, these are some **minimum figures** with which to work:

Minimum daily allocation (l/un.)	
Urban resident	50
Rural resident	30
Student	5
Outpatient	5
Inpatient	60
Ablution	2
Camel (once a week)	250
Goat and sheep	5
Cow	20
Horses, mules, and donkeys	20

In practice, the tendency is to provide the maximum amount of water such that:

- environmental health problems are avoided (stagnant water, over exploitation of the source)
- people are prepared to pay
- the cost is appropriate to the local economic conditions.

Work out the <u>minimum</u> base demand for a town of 1,300 people, where the average family has 5 members, 2 goats and one cow.

1. To work out the number of animals, you need to work out the approximate number of families in the town:

   1,300 people / 5 people/family = 260 families.

2. The number of animals is:

   260 families * 2 goats/family = 520 goats
   260 families * 1 cow/family = 260 cows

3. The base demand is:

   1,300 people * 30 l/person = 39,000 litres
   560 goats * 5 l/goat = 2,800 litres
   260 cows * 20 l/cow = 5,200 litres

   47,000 litres per day.

What is the flow required to supply the tank of a hospital centre which has 35 beds, and which treats 410 people every day? What if there was also a maternity ward with 8 births every day?

1. The minimum base demand would be:

   35 beds * 60 l/inpatient = 2,100 litres
   410 people treated * 5 l/outpatient = 2,050 litres

   4,150 litres per day.

2. The flow would be 4,150 litres in 24 hours. Coverting the units to litres per second:

   4,150 l/day * (1 day /24h) * (1h /3,600s) = 0.048 l/s

3. It´s always a good idea to check the demand estimates. In the case of a birth, we can´t be sure how much water will be consumed. The onsite doctor can give us an idea: 50 litres/birth.

Always check that the figures you are quoted make sense. A doctor in Mauritania may claim to use 5 litres, but this is really incompatible with decent hygiene during childbirth. It doesn´t provide enough water to wash the mother, the baby, the equipment, and the delivery room.

3. 2 FUTURE PROJECTIONS

Populations grow. If you design a system without taking this into account, it´ll soon be overtaken by rising demand. Expanding a system is much more expensive than building it with sufficient capacity from the start. In general, 30 years is the standard design period, although this can change depending on the circumstances.

To try and work out the future projection, there are two basic methods:

Geometric projection

Usually census data is used together with a growth rate. The projected population in the future is:

$$P_f = P_o \left(1 + \frac{i}{100} \right)^t$$

P_f , future population
P_o , current population
i , growth rate in %
t , time in years

You then work just with the future population figure rather than with the current, so as to take into account it´s growth.

Saturation

Imagine a beach early in the morning. As the day goes by, it starts filling up with people. There comes a certain point where new people arriving decide to go somewhere else when they see the beach is too full. This is how a balance is established in relation to peak density, for example, 1 person per meter squared. If the beach is 1,000 square meters, when it fills up in the future, it will have 1,000 people.

This focus is very useful in some cases, such as urban populations with rapid and unpredictable settlements.

What will the base demand be in 30 years for the population of exercise 21 that grows at a rate of 2% annually?

1. The population 30 years later is:

$$P_f = P_o\left(1 + \frac{i}{100}\right)^t \qquad P_f = 1{,}300\left(1 + \frac{2}{100}\right)^{30}$$

P_{30} = 2,355 people

2. Assuming the average family size does not change:
   2,355 people / 5 people/family = 471 families.

3. Assuming the number of animals does not change:
   471 families * 2 goats/family = 942 goats
   471 families * 1 cow/family = 471 cows

4. Assuming the personal demand does not change:

   2,355 people * 30 l/person = 70,650 litres
   942 goats * 5 l/goat = 4,710 litres
   471 cow * 20 l/cow = 9,420 litres

   84,780 litres per day.

Notice how there are many *assumptions* here. Future projections are not a precise science, they are an approximated prediction, and predictions are sometimes...well...have a look at this:

➢ ***"The market for photocopying machines is, at most, 5,000 globally."***

(IBM to the future founders of Xerox, 1969)

If in 1,995 a census showed a settlement had 10,000 inhabitants, and in 2,005 showed it has 12,000, what would be the population forecast if you were designing a system for 2,040?

1. Let´s work out the growth rate *i*:

$i = (P_{2,005} - P_{1,995}) / (P_{1,995} * t) = 100 * (12,000 - 10,000) / (10,000 * 10) = 2\%$

2. Let´s apply the formula starting with 2,005:

$$P_f = P_o \left(1 + \frac{i}{100}\right)^t \quad P_f = 12,000 \left(1 + \frac{2}{100}\right)^{35}$$

P_{2040} = 23,999 people.

An area of 1.3km^2 in a city is being rapidly populated with immigration from rural areas. Before conditions get worse, basic services need to be provided. In other areas of the same city which have been populated by rural immigration, each family occupies approximately 260m^2 and averages 7 people. What is the population to be supplied?

1. The peak density is:

7 people / 260 m^2 * 1,000,000 m^2 / 1 km^2 = 26,923 people / km^2

2. The population will be: 1.3 km^2 * 26,923 people / km^2 = 35,000 people

3. 3 MAXIMUM DESIGN FLOW

Very briefly, there are 3 ways to determine the design flow, depending on the situation:

1. **All taps open.** A tap is assigned with a specific number of users, and the flow is designed to supply all taps simultaneously. This is the approach taken in emergencies, refugee camps, and similar situations. For example, one tap is assigned 250 people with a flow of 0.2 l/s.

2. **Temporal variations.** Even more important than knowing *how* much water is consumed, is knowing *when* it will be consumed. People use more water at certain times of the day, on certain days of the week, and sometimes during certain months of the year. To take this into account, a multiplier is used to increase the average flow by a certain factor.

   The idea is to design the system for the worst case scenario: "the hour of highest demand, the day of highest demand, and the month of highest demand." This is the approach used most often for larger populations. For a gravity flow water project calculated by hand, multiply the average demand by a factor of between 3.5 and 4.5 (3.5 if the demand is more regular and the hot season is less pronounced etc).

It´s a good idea to read section 2.5 of the theory book to get a better understanding of the nature of these variations.

3. **Simultaneity.** Use this approach for systems and <u>parts of systems</u> which have less than 250 connections. Nevertheless, for large systems it´s good working practice to establish a minimum pipe diameter (generally 75mm to 100mm). This diameter is sufficient to not have to take into account simultaniety, as it can supply more than 250 connections. A minimum diameter is established to make enlargements easier, avoid blockages and provide a certain amount of protection against fires.

Avoid installing pipes which are too small, especially when the distances are large. Putting in a 40mm pipe over 2km for example, is a really bad idea. It´ll get blocked, and it then becomes very difficult to work out where the blockage is. Be careful with saving money here! It´ll soon end up becoming very expensive for the local population.

To work out the design flow for simultaneity, use this graph to find the multiplier for the average demand (Arizmendi 1,991):

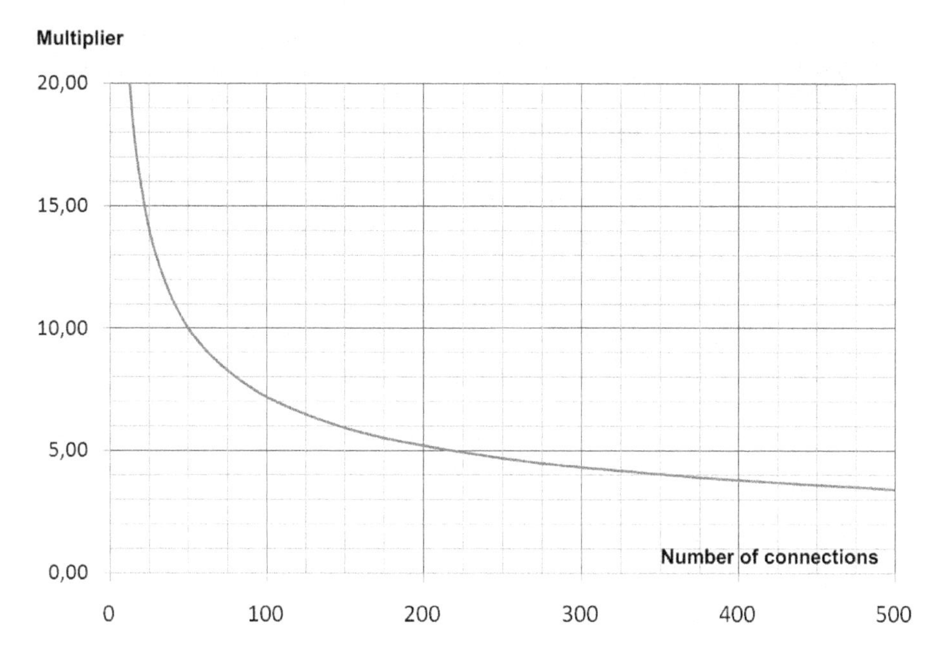

Most of the time, choosing the method for calculating the corrected demand comes down to 3 basic questions, which you can see in the flow diagram below:

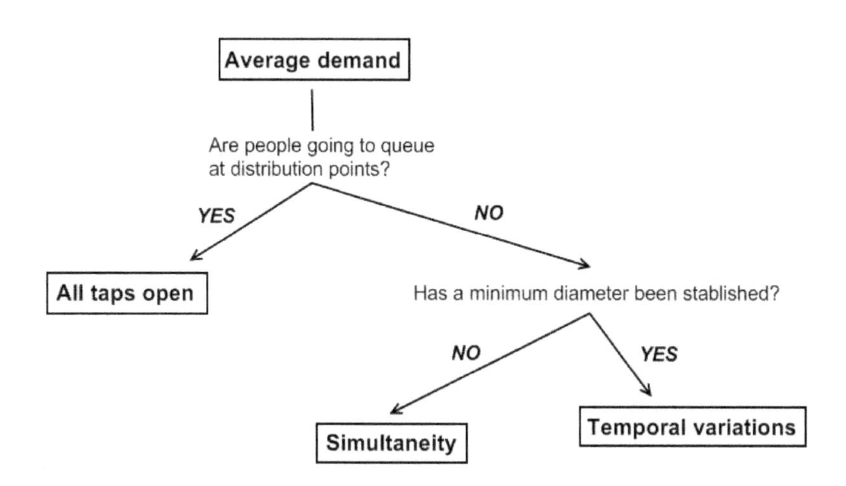

Following an earthquake, a refugee camp is being planned on flat terrain for 5,000 people. The site is arable land 1.4km x 3.5km with an existing irrigation system, which supplies a central point, 0, with a pressure of 3 bar and no flow limitations. Calculate the required system.

1. This is an emergency system. We can use the minimum standards from Chapter 2 of the Sphere Project: www.sphereproject.org

Since the quantity of water is not a limitation, we´re only concerned with 3 standards:

✓ The maximum distance between any home and the nearest water supply point is 500 meters.

✓ It doesn´t take more than 3 minutes to fill a container of 20 litres.

✓ Maximum of 250 people per tap.

For practical purposes:

2. The first standard requires that each tap be at a maximum distance of 700 meters from one another[2] if the layout is regular. One way of doing it would be this:

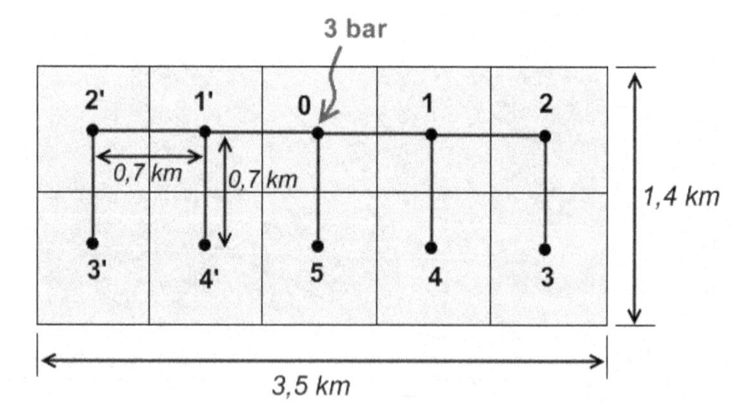

[2] In a regular layout, the simplest is to draw out boxes inside circles with a radius of 500m. Applying Pythagoras´ theorem, each side comes out as $2r/\sqrt{2}$=707.1m. Don´t worry if you don´t understand the reasoning for now, it´s not really that important in the context of this book.

3. The second requires a minimum flow of:

Q=(20 litres / 3 min) * (1 min / 60s) = 0.11 l/s.

This is a really small flow. 0.2 l/s is more appropriate.

4. To meet the third criteria, you have to work out how many people there are in the distribution area of each public tap stand:

5,000 people / 10 zones = 500 people/zone.

If one tap for every 250 people is needed, a minimum of 2 taps will be necessary. This means the flow at each supply point will be 3 * 0.2 l/s = 0.6 l/s.

5. To calculate the design flow you need to look at the context. As it´s a refugee camp:

- Intuitively the timescale is short → No future projections are made
- There´ll be queues at each tap stand, which means they´ll be used simultaneously → Calculate for all taps open

Now you´re ready to do the calculation. Note that the system is symmetrical, which means you only have to size the pipes which supply nodes 1, 2, 3, 4 y 5. The pipes that go from 1', 2', 3' y 4' will be the same as the ones that go from 1, 2, 3, 4.

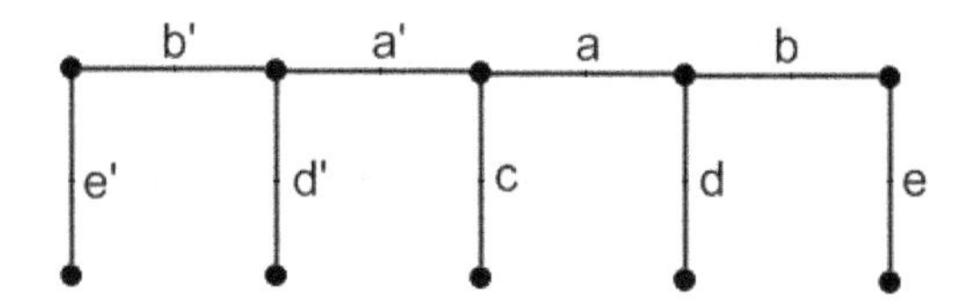

6. Determine the flows each pipe will carry if the supply points have a demand of 0.6 l/s. Remember the 0 node doesn´t affect the calculation because you are working with a supply line which has no flow limitations:

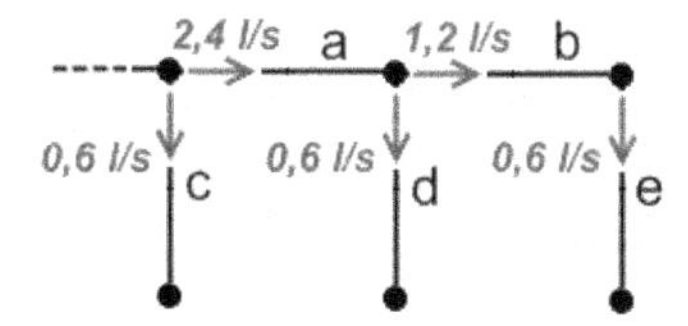

7. The system is flat and begins with 3 bar of pressure. The longest route, which goes from 0 to 3 has 3 * 0.7 km = 2.1 km. If you aim to get there with 15m of pressure, the head loss is:

$$J = 15m / 2.1\ km = 7.14\ m$$

The material to use is HDPE, as it´s quick to assemble (it comes in rolls), it´s strong, and because PVC is damaged when exposed to the sun (in an emergency it´s best not to wait for the pipes to be buried).

The pipes which are closest to the required head loss for 2.4 l/s, 1.2 l/s and 0.6 l/s are:

HDPE 90 mm, $J_{2.4\ l/s}$ = 3.5 m/km → Pipes *a* and *a'*.
HDPE 63 mm, $J_{1.2\ l/s}$ = 6 m/km → Pipes *b* y *b'*.
HDPE 40 mm, $J_{0.6\ l/s}$ = 15 m/km.

This pipe would be acceptable, but seeing as in an emergency you can´t kick people out because the 5000 person limit has been exceeded, it´s best to design in some extra carrying capacity. Pipes *e* and *e'* will also be installed in 63mm.

HDPE 63 mm, $J_{0.6\ l/s}$ = 1.7 m/km

8. The pressure at the different points is:

P_0 = 30 m
P_1 = 30 m – 0.7 km * 3.5 m/km = 27.55m
P_2 = 27.55m – 0.7 km * 6 m/km = 23.35m
P_3 = 23.35m – 0.7 km * 1.7 m/km = 22.16m

9. The pipes *c* and *d* can be 40mm:

P_5 = 30m – 0.7 km * 15 m/km = 19.5m
P_4 = 27.55m – 0.7 km * 15 m/km = 17.05m

The system looks this:

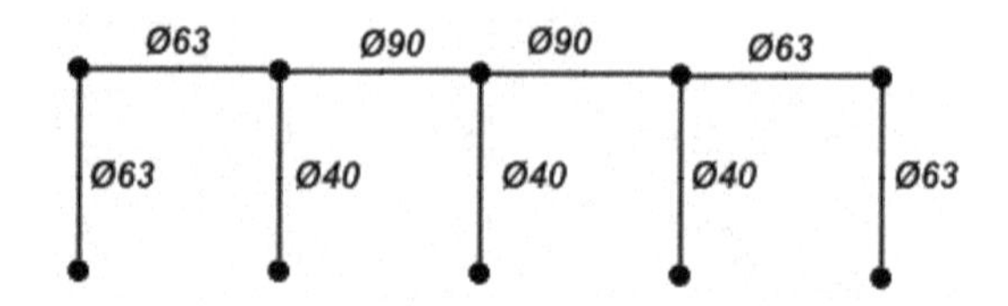

A town of 9600 people grows at 3% annually. The sources which supplied it can no longer meet the growing demand, and a new pipeline is being planned to draw water from a reservoir, 8km away at an elevation of 65m, descending to 21m in the town. The slope is uniform. The point of connection must have a minimum residual pressure of 1.5 bar. The monthly bills show a demand of 90 l/person. The habits and customs of the inhabitants' don´t vary much one from the other.

1. The population forecast for 30 years will be:

$$P_f = P_o \left(1 + \frac{i}{100}\right)^t \qquad P_f = 9{,}600\left(1 + \frac{3}{100}\right)^{30}$$

   P_f = 23,300 people.

2. As there are more than 200 connections, the temporal variations approach is used. If the inhabitants have similar habits they´ll use water at similar times. This puts pressure on the system as demand will be concentrated during these times. We´ll use the multiplier of 4.2, between the range of 3.5 and 4.5 (that´s right, an educated guess! The data for more precise calculations aren´t available.)

   Q = 23,300 people * 90 l/people days * 4.2 = 8,807,400 l/day
   Q = 8,807,400 l/day * 1 day / 86,400 s = 102 l/s

3. The maximum head loss is:

   J_{max} = (65m – 21m -15m) / 8 km = 3.62 m/km.

A topographic survey is always done. Even if the slope appears homogenous, the surveys can throw up some surprises. No project should be without a topographic survey, it´s only omitted here to make the exercise more manageable.

4. Let´s look at the tables. Seeing as in the book we´re short of tables, let´s look at the online version:

   www.arnalich.com/dwnl/headloss.zip

HDPE pipes larger than 8” require expensive welding equipment. If there are none available in the region, it may not make sense to invest in one, as the project may end up being too costly.

5. We´ll go for PVC:

J_{315} = 6 m/km and J_{400} = 1.9 m/km

Large diameter pipes are very expensive and it´s worth adjusting the diameters precisely to reduce costs.

6. A combination of pipe sizes will be installed:

J_{315}*x + J_{400} (d-x) = D

6 m/km * x + 1.9 m/km (8 -x) = 65m – 21m -15m

4.1x + 15.2 = 29

4.1x = 13.8 → x = 3,366m of 315mm pipe

d-x = 8,000m – 3,366m = 4,634m of 400mm pipe.

From a tank at an elevation of 18m above a flat terrain, the following system needs to be supplied for a small group of villages with a demand of 40 litres/day*person. There is a connection for every 8 people. In village A there are 25 connections, 100 in B, and 200 in C; for a while they have not grown:

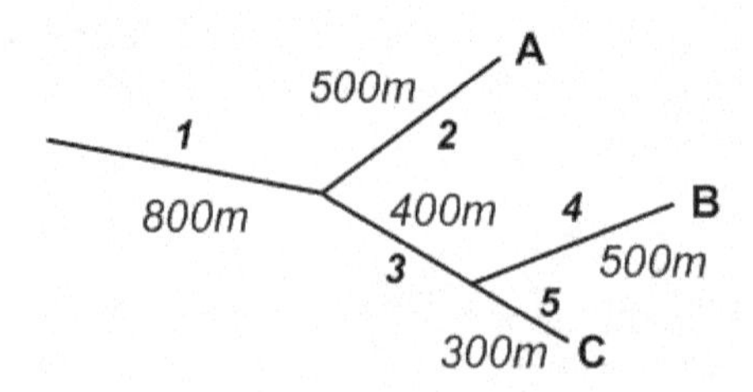

1. If they don´t grow, there´s no need to project into the future. On the other hand, the system is small and the distances are short. There´ll be no minimum diameter, and seeing as the connection groups are small, the simultaneity approach will be used.

2. Just as the flow was divided up earlier, it´s best to divide up the connections for each pipeline:

- Pipe 1, 325 connections → No simultaneity needed. You can use the graph or use the multiplier of 3.5-4.5. Note that the results are very similar. We´ll take the temporal variation multiplier to be C_t =4.
- Pipe 2, 25 connections → Simultaneity, coefficient C_2 = 14.
- Pipe 3, 300 connections → No simultaneity needed.
- Pipe 4, 100 connections → Simultaneity, coefficient C_4 = 7.2.
- Pipe 5, 200 connections → Simultaneity, coefficient C_4 = 5.3.

The values of simultaneity are obtained from the graph:

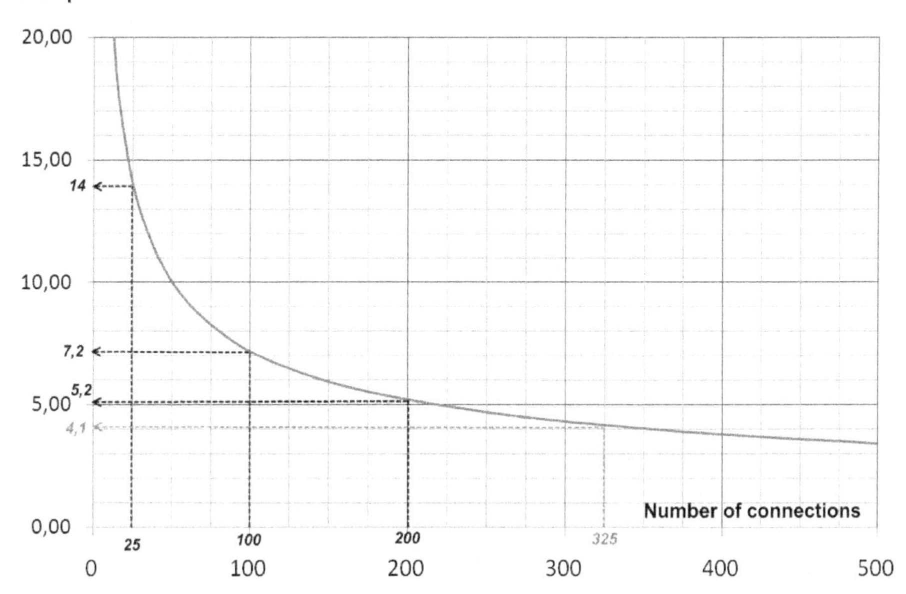

3. The corrected flows of each pipe are calculated, multiplying the average flow by the multiplier. The average flow for each connection is:

Q= 40 l/per*day * 8 per/ connection * (1 day / 86,400s) = 0.003704 l/s

For example, pipe 1 has an average flow of 325 connections * 0.003704 l/s*connection = 1.204 l/s. The adjusted flow is 1.204 l/s * 4 = 4.82 l/s. And so on for each pipe. The results can be seen in the following table:

0.003704	Connections	Q aver.	C. simult.	Q. adjusted
Pipe 1	325	1.204	4	4.82
Pipe 2	25	0.093	14	1.30
Pipe 3	300	1.111	4	4.44
Pipe 4	100	0.370	7.2	2.67
Pipe 5	200	0.741	5.2	3.85

4. The diameters needed are calculated with the adjusted flows. The longest route, pipes 1, 3 and 4, is 1700m. Aiming to get there with 10m:

$$J_{max} = (18m - 10m) / 1.7\ km = 4.7\ m/km$$

As there´s not much of a pressure margin, it´s best to choose pipes that have a head loss equal to or less than 4.7 m/km between them all, without much need of adjustment. In HDPE:

Pipe 1 will be 110mm → $J_{4.82\ l/s}$ = 4.75 m/km
Pipe 3 will be 110mm → $J_{4.44\ l/s}$ = 4.25 m/km
Pipe 4 will be 90mm → $J_{2.67\ l/s}$ = 4.25 m/km
Pipe 2 will be 90mm → $J_{1.3\ l/s}$ = 1.2 m/km
Pipe 5 will be 110mm → $J_{3.85\ l/s}$ = 3.25 m/km

I´ll leave it to you to check that the pressures are greater than 10m in all the points.

3. 4 UNACCOUNTED-FOR WATER

Unaccounted-for water (UFW) is a little box of disasters in which you´ll find system leaks, unauthorized connections, authorized but unpaid connections…In some systems it´s less a question of unmeasured as "unconsumed demand." For example, in an emergency system with public tap stands where the users are going to consume a minimum of 15 litres/day, you have to take into account that some of it will be wasted. If to this you add the standard losses from a new system of around 20%, you´ll probably end up with around 10 litres. With such small amounts of water, the difference between 10 and 15 litres is a world of difference.

Take something in the region of 20% more to take this into account, which means if the base demand was 10 l/s, the corrected demand will be 10* 1.2 = 12 l/s (adding 20% or multiplying by 1.2 is the same thing. 35% would mean multiplying by 1.35).

This diagram sums up the process of this chapter:

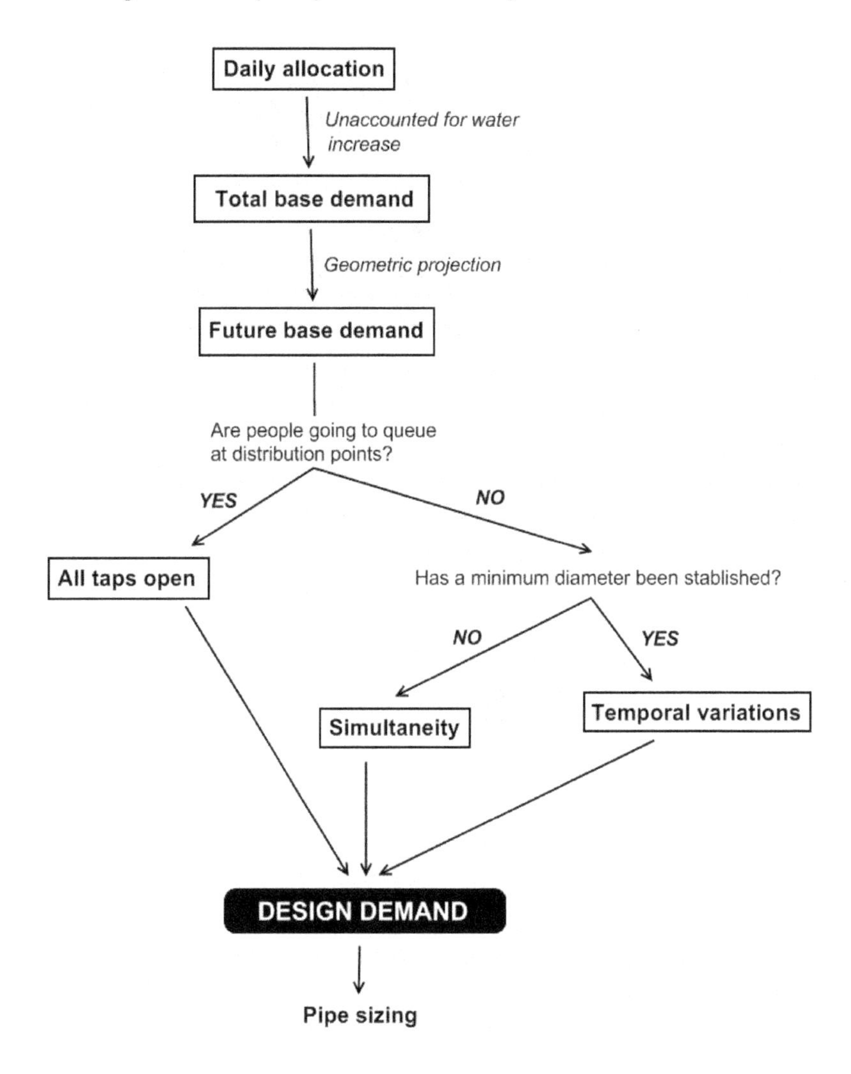

The order in which you do the multiplications doesn`t affect the result (adding 20% at the beginning or the end makes no difference).

About the author

Santiago Arnalich

At 26 years old, he began as the coordinator of the Kabul Project CAWWS Water Supply, providing water to 565,000 people, probably the most important water supply project to date. Since then, he has designed improvements for more than a million people, including refugee camps in Tanzania, the city of Meulaboh following the Tsunami, and the poor neighbourhoods of Santa Cruz, Bolivia.

Currently he is founder and coordinator of Arnalich, Water and Habitat, a private entity with a strong social commitment dedicated to promoting the impact of humanitarian organisations through training and technical assistance in the fields of drinking water supply and environmental engineering.

Bibliography

1. Arizmendi, L. (1,991). *Instalaciones Urbanas, Infraestructura y Planeamiento.* Librería Editorial Bellisco.

2. Arnalich, S. (2,008). *Gravity Flow Water Supply.* Arnalich, water and habitat.

   www.arnalich.com/en/libros.html

3. Arnalich, S. (2,007). *Epanet in Aid: How to calculate a water network by computer.* Arnalich, water and habitat.

   www.arnalich.com/en/libros.html

4. Arnalich, S. (2,007*) Epanet in Aid. 44 progressive exercises to calculate water networks by computer.* Arnalich, water and habitat.

   www.arnalich.com/en/libros.html

5. Department of Lands, Valuation and Water (1,983). *Gravity Fed Rural Piped Water Schemes.* Republic of Malawi.

6. Fuertes, V. S. et al (2,002). *Modelación y Diseño de Redes de Abastecimiento de Agua.* Servicio de Publicación de la Universidad Politécnica de Valencia.

7. Jordan T. D. (1,980). *A Handbook of Gravity-Flow Water Systems.* Intermediate Technology Publications.

8. Mays L. W. (1,999). *Water Distribution Systems Handbook.* McGraw-Hill Press.

9. Santosh Kumar Garg (2,003). *Water Supply Engineering.* 14º ed. Khanna Publishers.

10. Stephenson, D. (1,981). *"Pipeline Design for Water Engineers".* Ed. Elsevier.

11. Walski, T. M. y otros (2,003). *Advanced Water Distribution Modeling and Management.* Haestad Press, USA. Haestad methods.

    http://www.haestad.com/library/books/awdm/online/wwhelp/wwhimpl/js/html/wwhelp.htm

APPENDICES

A. FRICTION LOSS TABLES. PLASTIC PIPE

(Courtesy of Uralita)

Below you will find the friction loss tables for the most commonly used pipes. Due to space limitations not all pipes are listed. If you´re looking for data that isn´t available here, go to www.arnalich.com/dwnl/headloss.zip.

To use the tables you need to know what material you´re working with, the maximum pressure, and the type of water you´ll be transporting (clean/dirty). For a flow of 0.02 l/s, an HDPE of 25mm at 16 bar, carrying clean water (k=0.01), has a head loss of 0.6 m/km.

25 - PN 16 **CLEAN WATER: K=0.01**

Head loss (m/km)	Q (l/s)	V (m/s)
0.50	0.018	0.06
0.60	0.020	0.06
0.70	0.022	0.07

J, head loss; Q, flow and V, velocity.

Important: head loss varies somewhat from one manufacturer to another. If a manufacturer provides you with reliable data, use this instead.

Metal pipe is specified with the internal diameter. Plastic pipe with the external diameter. This table shows the approximate internal diameters (ID) for plastic pipe:

ND	25	32	40	50	63	75	90	110	125	140	160	180	200	250	315	400
ID HDPE	20	26	35	44	55	66	79	97	110	123	141	159	176	220	277	353
ID PVC	21	29	36	45	57	68	81	102	115	129	148	159	185	231	291	369

HDPE 25 -ID 20.4mm- PN 16		
J (m/km)	Q (l/s)	v (m/s)
0.50	0.018	0.06
0.60	0.020	0.06
0.70	0.022	0.07
0.80	0.024	0.07
0.90	0.026	0.08
1.00	0.028	0.08
1.10	0.029	0.09
1.20	0.031	0.09
1.30	0.032	0.10
1.40	0.034	0.10
1.50	0.035	0.11
1.60	0.037	0.11
1.70	0.038	0.12
1.80	0.039	0.12
1.90	0.041	0.12
2.00	0.042	0.13
2.25	0.045	0.14
2.50	0.048	0.15
2.75	0.050	0.15
3.00	0.053	0.16
3.25	0.056	0.17
3.50	0.058	0.18
3.75	0.061	0.19
4.00	0.063	0.19
4.25	0.065	0.20
4.50	0.067	0.21
4.75	0.070	0.21
5.00	0.072	0.22
5.50	0.076	0.23
6.00	0.080	0.24
6.50	0.084	0.26
7.00	0.087	0.27
7.50	0.091	0.28
8.00	0.094	0.29
8.50	0.098	0.30
9.00	0.101	0.31
10.00	0.107	0.33
12.00	0.119	0.36
15.00	0.136	0.41
20.00	0.160	0.49
30.00	0.202	0.62
45.00	0.254	0.78
60.00	0.299	0.91

HDPE 32 - ID 26.2mm- PN 16		
J (m/km)	Q (l/s)	v (m/s)
0.50	0.037	0.07
0.60	0.041	0.08
0.70	0.045	0.08
0.80	0.049	0.09
0.90	0.052	0.10
1.00	0.056	0.10
1.10	0.059	0.11
1.20	0.062	0.11
1.30	0.065	0.12
1.40	0.068	0.13
1.50	0.071	0.13
1.60	0.073	0.14
1.70	0.076	0.14
1.80	0.079	0.15
1.90	0.081	0.15
2.00	0.084	0.16
2.25	0.090	0.17
2.50	0.095	0.18
2.75	0.101	0.19
3.00	0.106	0.20
3.25	0.111	0.21
3.50	0.116	0.22
3.75	0.121	0.22
4.00	0.125	0.23
4.25	0.130	0.24
4.50	0.134	0.25
4.75	0.139	0.26
5.00	0.143	0.26
5.50	0.151	0.28
6.00	0.159	0.29
6.50	0.166	0.31
7.00	0.173	0.32
7.50	0.180	0.33
8.00	0.187	0.35
8.50	0.194	0.36
9.00	0.200	0.37
10.00	0.213	0.39
12.00	0.236	0.44
15.00	0.269	0.50
20.00	0.316	0.59
30.00	0.398	0.74
45.00	0.501	0.93
60.00	0.589	1.09

HDPE 40 - ID 35.2mm- PN 10		
J (m/km)	Q (l/s)	v (m/s)
0.50	0.084	0.09
0.60	0.093	0.10
0.70	0.102	0.10
0.80	0.111	0.11
0.90	0.118	0.12
1.00	0.126	0.13
1.10	0.133	0.14
1.20	0.140	0.14
1.30	0.147	0.15
1.40	0.153	0.16
1.50	0.160	0.16
1.60	0.166	0.17
1.70	0.172	0.18
1.80	0.178	0.18
1.90	0.183	0.19
2.00	0.189	0.19
2.25	0.202	0.21
2.50	0.215	0.22
2.75	0.227	0.23
3.00	0.239	0.25
3.25	0.250	0.26
3.50	0.261	0.27
3.75	0.272	0.28
4.00	0.282	0.29
4.25	0.292	0.30
4.50	0.302	0.31
4.75	0.311	0.32
5.00	0.320	0.33
5.50	0.338	0.35
6.00	0.356	0.37
6.50	0.372	0.38
7.00	0.388	0.40
7.50	0.404	0.42
8.00	0.419	0.43
8.50	0.434	0.45
9.00	0.448	0.46
10.00	0.476	0.49
12.00	0.528	0.54
15.00	0.599	0.62
20.00	0.705	0.72
30.00	0.885	0.91
45.00	1.111	1.14
60.00	1.304	1.34

HDPE 40 - ID 32.6mm- PN 16		
J (m/km)	Q (l/s)	v (m/s)
0.50	0.068	0.08
0.60	0.075	0.09
0.70	0.083	0.10
0.80	0.089	0.11
0.90	0.096	0.11
1.00	0.102	0.12
1.10	0.108	0.13
1.20	0.113	0.14
1.30	0.119	0.14
1.40	0.124	0.15
1.50	0.129	0.15
1.60	0.134	0.16
1.70	0.139	0.17
1.80	0.144	0.17
1.90	0.148	0.18
2.00	0.153	0.18
2.25	0.164	0.20
2.50	0.174	0.21
2.75	0.184	0.22
3.00	0.193	0.23
3.25	0.203	0.24
3.50	0.211	0.25
3.75	0.220	0.26
4.00	0.228	0.27
4.25	0.237	0.28
4.50	0.244	0.29
4.75	0.252	0.30
5.00	0.260	0.31
5.50	0.274	0.33
6.00	0.288	0.35
6.50	0.302	0.36
7.00	0.315	0.38
7.50	0.328	0.39
8.00	0.340	0.41
8.50	0.352	0.42
9.00	0.364	0.44
10.00	0.386	0.46
12.00	0.429	0.51
15.00	0.487	0.58
20.00	0.573	0.69
30.00	0.720	0.86
45.00	0.904	1.08
60.00	1.061	1.27

HDPE 63 - ID 55.4mm- PN 10		
J (m/km)	Q (l/s)	v (m/s)
0.50	0.293	0.12
0.60	0.326	0.14
0.70	0.357	0.15
0.80	0.385	0.16
0.90	0.412	0.17
1.00	0.438	0.18
1.10	0.463	0.19
1.20	0.487	0.20
1.30	0.510	0.21
1.40	0.532	0.22
1.50	0.553	0.23
1.60	0.574	0.24
1.70	0.594	0.25
1.80	0.614	0.25
1.90	0.633	0.26
2.00	0.652	0.27
2.25	0.698	0.29
2.50	0.741	0.31
2.75	0.782	0.32
3.00	0.822	0.34
3.25	0.860	0.36
3.50	0.897	0.37
3.75	0.933	0.39
4.00	0.968	0.40
4.25	1.002	0.42
4.50	1.035	0.43
4.75	1.067	0.44
5.00	1.099	0.46
5.50	1.159	0.48
6.00	1.218	0.51
6.50	1.274	0.53
7.00	1.329	0.55
7.50	1.381	0.57
8.00	1.432	0.59
8.50	1.482	0.61
9.00	1.531	0.63
10.00	1.624	0.67
12.00	1.799	0.75
15.00	2.038	0.85
20.00	2.393	0.99
30.00	2.998	1.24
45.00	3.752	1.56
60.00	4.396	1.82

HDPE 63 - ID 51.4mm- PN 16		
J (m/km)	Q (l/s)	v (m/s)
0.50	0.239	0.12
0.60	0.265	0.13
0.70	0.290	0.14
0.80	0.314	0.15
0.90	0.336	0.16
1.00	0.357	0.17
1.10	0.377	0.18
1.20	0.396	0.19
1.30	0.415	0.20
1.40	0.433	0.21
1.50	0.451	0.22
1.60	0.468	0.23
1.70	0.484	0.23
1.80	0.501	0.24
1.90	0.516	0.25
2.00	0.532	0.26
2.25	0.569	0.27
2.50	0.604	0.29
2.75	0.638	0.31
3.00	0.671	0.32
3.25	0.702	0.34
3.50	0.732	0.35
3.75	0.762	0.37
4.00	0.790	0.38
4.25	0.818	0.39
4.50	0.845	0.41
4.75	0.871	0.42
5.00	0.897	0.43
5.50	0.947	0.46
6.00	0.994	0.48
6.50	1.040	0.50
7.00	1.085	0.52
7.50	1.128	0.54
8.00	1.170	0.56
8.50	1.211	0.58
9.00	1.250	0.60
10.00	1.327	0.64
12.00	1.470	0.71
15.00	1.666	0.80
20.00	1.957	0.94
30.00	2.452	1.18
45.00	3.070	1.48
60.00	3.598	1.73

HDPE 90 - ID 79.2mm- PN 10		
J (m/km)	Q (l/s)	v (m/s)
0.50	0.780	0.16
0.60	0.866	0.18
0.70	0.946	0.19
0.80	1.021	0.21
0.90	1.092	0.22
1.00	1.160	0.24
1.10	1.225	0.25
1.20	1.287	0.26
1.30	1.347	0.27
1.40	1.405	0.29
1.50	1.461	0.30
1.60	1.516	0.31
1.70	1.569	0.32
1.80	1.620	0.33
1.90	1.671	0.34
2.00	1.720	0.35
2.25	1.839	0.37
2.50	1.951	0.40
2.75	2.059	0.42
3.00	2.163	0.44
3.25	2.263	0.46
3.50	2.359	0.48
3.75	2.452	0.50
4.00	2.543	0.52
4.25	2.631	0.53
4.50	2.717	0.55
4.75	2.801	0.57
5.00	2.882	0.59
5.50	3.041	0.62
6.00	3.192	0.65
6.50	3.339	0.68
7.00	3.480	0.71
7.50	3.617	0.73
8.00	3.749	0.76
8.50	3.878	0.79
9.00	4.004	0.81
10.00	4.246	0.86
12.00	4.699	0.95
15.00	5.318	1.08
20.00	6.236	1.27
30.00	7.798	1.58
45.00	9.740	1.98
60.00	11.398	2.31

HDPE 90 - ID 73.6mm- PN 16		
J (m/km)	Q (l/s)	v (m/s)
0.50	0.639	0.15
0.60	0.709	0.17
0.70	0.775	0.18
0.80	0.837	0.20
0.90	0.895	0.21
1.00	0.950	0.22
1.10	1.004	0.24
1.20	1.055	0.25
1.30	1.104	0.26
1.40	1.152	0.27
1.50	1.198	0.28
1.60	1.243	0.29
1.70	1.286	0.30
1.80	1.329	0.31
1.90	1.370	0.32
2.00	1.410	0.33
2.25	1.508	0.35
2.50	1.600	0.38
2.75	1.689	0.40
3.00	1.774	0.42
3.25	1.856	0.44
3.50	1.936	0.45
3.75	2.012	0.47
4.00	2.087	0.49
4.25	2.159	0.51
4.50	2.230	0.52
4.75	2.299	0.54
5.00	2.366	0.56
5.50	2.496	0.59
6.00	2.621	0.62
6.50	2.741	0.64
7.00	2.857	0.67
7.50	2.970	0.70
8.00	3.079	0.72
8.50	3.185	0.75
9.00	3.288	0.77
10.00	3.487	0.82
12.00	3.860	0.91
15.00	4.370	1.03
20.00	5.125	1.20
30.00	6.411	1.51
45.00	8.011	1.88
60.00	9.377	2.20

HDPE 110 - ID 96.8mm- PN 10		
J (m/km)	Q (l/s)	v (m/s)
0.50	1.347	0.18
0.60	1.495	0.20
0.70	1.632	0.22
0.80	1.761	0.24
0.90	1.883	0.26
1.00	1.999	0.27
1.10	2.110	0.29
1.20	2.216	0.30
1.30	2.319	0.32
1.40	2.418	0.33
1.50	2.514	0.34
1.60	2.608	0.35
1.70	2.698	0.37
1.80	2.787	0.38
1.90	2.873	0.39
2.00	2.957	0.40
2.25	3.160	0.43
2.50	3.353	0.46
2.75	3.537	0.48
3.00	3.714	0.50
3.25	3.885	0.53
3.50	4.049	0.55
3.75	4.209	0.57
4.00	4.364	0.59
4.25	4.514	0.61
4.50	4.661	0.63
4.75	4.804	0.65
5.00	4.943	0.67
5.50	5.213	0.71
6.00	5.472	0.74
6.50	5.722	0.78
7.00	5.962	0.81
7.50	6.196	0.84
8.00	6.422	0.87
8.50	6.642	0.90
9.00	6.856	0.93
10.00	7.268	0.99
12.00	8.040	1.09
15.00	9.095	1.24
20.00	10.656	1.45
30.00	13.312	1.81
45.00	16.612	2.26
60.00	19.426	2.64

HDPE 110 - ID 90mm- PN 16		
J (m/km)	Q (l/s)	v (m/s)
0.50	1.105	0.17
0.60	1.227	0.19
0.70	1.339	0.21
0.80	1.445	0.23
0.90	1.546	0.24
1.00	1.641	0.26
1.10	1.732	0.27
1.20	1.820	0.29
1.30	1.904	0.30
1.40	1.986	0.31
1.50	2.065	0.32
1.60	2.142	0.34
1.70	2.217	0.35
1.80	2.289	0.36
1.90	2.360	0.37
2.00	2.430	0.38
2.25	2.597	0.41
2.50	2.755	0.43
2.75	2.907	0.46
3.00	3.053	0.48
3.25	3.193	0.50
3.50	3.329	0.52
3.75	3.460	0.54
4.00	3.588	0.56
4.25	3.712	0.58
4.50	3.832	0.60
4.75	3.950	0.62
5.00	4.065	0.64
5.50	4.287	0.67
6.00	4.501	0.71
6.50	4.706	0.74
7.00	4.905	0.77
7.50	5.097	0.80
8.00	5.283	0.83
8.50	5.464	0.86
9.00	5.641	0.89
10.00	5.981	0.94
12.00	6.617	1.04
15.00	7.486	1.18
20.00	8.774	1.38
30.00	10.965	1.72
45.00	13.688	2.15
60.00	16.010	2.52

HDPE 160 - ID 141mm- PN 10		
J (m/km)	Q (l/s)	v (m/s)
0.50	3.732	0.24
0.60	4.136	0.26
0.70	4.512	0.29
0.80	4.865	0.31
0.90	5.198	0.33
1.00	5.515	0.35
1.10	5.818	0.37
1.20	6.110	0.39
1.30	6.390	0.41
1.40	6.661	0.43
1.50	6.923	0.44
1.60	7.178	0.46
1.70	7.426	0.48
1.80	7.667	0.49
1.90	7.902	0.51
2.00	8.131	0.52
2.25	8.684	0.56
2.50	9.209	0.59
2.75	9.711	0.62
3.00	10.193	0.65
3.25	10.656	0.68
3.50	11.104	0.71
3.75	11.538	0.74
4.00	11.959	0.77
4.25	12.367	0.79
4.50	12.765	0.82
4.75	13.154	0.84
5.00	13.532	0.87
5.50	14.265	0.91
6.00	14.968	0.96
6.50	15.644	1.00
7.00	16.298	1.04
7.50	16.930	1.08
8.00	17.543	1.12
8.50	18.138	1.16
9.00	18.718	1.20
10.00	19.835	1.27
12.00	21.924	1.40
15.00	24.775	1.59
20.00	28.994	1.86
30.00	36.156	2.32
45.00	45.043	2.88
60.00	52.609	3.37

HDPE 160 - ID 130.8mm- PN 16		
J (m/km)	Q (l/s)	v (m/s)
0.50	3.046	0.23
0.60	3.377	0.25
0.70	3.685	0.27
0.80	3.973	0.30
0.90	4.246	0.32
1.00	4.506	0.34
1.10	4.754	0.35
1.20	4.992	0.37
1.30	5.222	0.39
1.40	5.443	0.41
1.50	5.658	0.42
1.60	5.867	0.44
1.70	6.069	0.45
1.80	6.267	0.47
1.90	6.459	0.48
2.00	6.647	0.49
2.25	7.100	0.53
2.50	7.530	0.56
2.75	7.941	0.59
3.00	8.335	0.62
3.25	8.715	0.65
3.50	9.082	0.68
3.75	9.438	0.70
4.00	9.782	0.73
4.25	10.117	0.75
4.50	10.443	0.78
4.75	10.761	0.80
5.00	11.072	0.82
5.50	11.672	0.87
6.00	12.248	0.91
6.50	12.803	0.95
7.00	13.338	0.99
7.50	13.856	1.03
8.00	14.359	1.07
8.50	14.847	1.10
9.00	15.322	1.14
10.00	16.238	1.21
12.00	17.951	1.34
15.00	20.290	1.51
20.00	23.750	1.77
30.00	29.627	2.20
45.00	36.921	2.75
60.00	43.133	3.21

HDPE 200 - ID 176.2mm- PN 10		
J (m/km)	Q (l/s)	v (m/s)
0.50	6.805	0.28
0.60	7.539	0.31
0.70	8.221	0.34
0.80	8.860	0.36
0.90	9.463	0.39
1.00	10.038	0.41
1.10	10.587	0.43
1.20	11.114	0.46
1.30	11.621	0.48
1.40	12.112	0.50
1.50	12.586	0.52
1.60	13.047	0.54
1.70	13.495	0.55
1.80	13.931	0.57
1.90	14.356	0.59
2.00	14.771	0.61
2.25	15.769	0.65
2.50	16.719	0.69
2.75	17.625	0.72
3.00	18.496	0.76
3.25	19.333	0.79
3.50	20.142	0.83
3.75	20.925	0.86
4.00	21.684	0.89
4.25	22.422	0.92
4.50	23.140	0.95
4.75	23.840	0.98
5.00	24.524	1.01
5.50	25.846	1.06
6.00	27.113	1.11
6.50	28.333	1.16
7.00	29.510	1.21
7.50	30.650	1.26
8.00	31.754	1.30
8.50	32.828	1.35
9.00	33.872	1.39
10.00	35.884	1.47
12.00	39.645	1.63
15.00	44.778	1.84
20.00	52.366	2.15
30.00	65.239	2.68
45.00	81.195	3.33
60.00	94.771	3.89

HDPE 200 - ID 163.6mm- PN 16		
J (m/km)	Q (l/s)	v (m/s)
0.50	5.573	0.27
0.60	6.175	0.29
0.70	6.734	0.32
0.80	7.258	0.35
0.90	7.754	0.37
1.00	8.225	0.39
1.10	8.676	0.41
1.20	9.108	0.43
1.30	9.525	0.45
1.40	9.928	0.47
1.50	10.317	0.49
1.60	10.695	0.51
1.70	11.063	0.53
1.80	11.421	0.54
1.90	11.770	0.56
2.00	12.111	0.58
2.25	12.931	0.62
2.50	13.710	0.65
2.75	14.455	0.69
3.00	15.170	0.72
3.25	15.858	0.75
3.50	16.523	0.79
3.75	17.166	0.82
4.00	17.790	0.85
4.25	18.396	0.88
4.50	18.986	0.90
4.75	19.562	0.93
5.00	20.123	0.96
5.50	21.209	1.01
6.00	22.251	1.06
6.50	23.254	1.11
7.00	24.221	1.15
7.50	25.158	1.20
8.00	26.066	1.24
8.50	26.948	1.28
9.00	27.807	1.32
10.00	29.461	1.40
12.00	32.554	1.55
15.00	36.775	1.75
20.00	43.017	2.05
30.00	53.609	2.55
45.00	66.741	3.17
60.00	77.918	3.71

PVC 40 - ID 36.2mm- PN 10		
J (m/km)	Q (l/s)	v m/s)
0.50	0.091	0.09
0.60	0.101	0.10
0.70	0.110	0.11
0.80	0.119	0.12
0.90	0.128	0.12
1.00	0.136	0.13
1.10	0.144	0.14
1.20	0.151	0.15
1.30	0.159	0.15
1.40	0.166	0.16
1.50	0.172	0.17
1.60	0.179	0.17
1.70	0.186	0.18
1.80	0.192	0.19
1.90	0.198	0.19
2.00	0.204	0.20
2.25	0.218	0.21
2.50	0.232	0.23
2.75	0.245	0.24
3.00	0.258	0.25
3.25	0.270	0.26
3.50	0.282	0.27
3.75	0.293	0.28
4.00	0.304	0.30
4.25	0.315	0.31
4.50	0.326	0.32
4.75	0.336	0.33
5.00	0.346	0.34
5.50	0.365	0.35
6.00	0.384	0.37
6.50	0.402	0.39
7.00	0.419	0.41
7.50	0.436	0.42
8.00	0.452	0.44
8.50	0.468	0.45
9.00	0.484	0.47
10.00	0.514	0.50
12.00	0.570	0.55
15.00	0.646	0.63
20.00	0.760	0.74
30.00	0.955	0.93
45.00	1.198	1.16
60.00	1.406	1.37

PVC 40 - ID 34mm- PN 16		
J (m/km)	Q (l/s)	v (m/s)
0.50	0.076	0.08
0.60	0.085	0.09
0.70	0.093	0.10
0.80	0.100	0.11
0.90	0.108	0.12
1.00	0.114	0.13
1.10	0.121	0.13
1.20	0.127	0.14
1.30	0.133	0.15
1.40	0.139	0.15
1.50	0.145	0.16
1.60	0.151	0.17
1.70	0.156	0.17
1.80	0.161	0.18
1.90	0.167	0.18
2.00	0.172	0.19
2.25	0.184	0.20
2.50	0.195	0.22
2.75	0.206	0.23
3.00	0.217	0.24
3.25	0.227	0.25
3.50	0.237	0.26
3.75	0.247	0.27
4.00	0.256	0.28
4.25	0.265	0.29
4.50	0.274	0.30
4.75	0.283	0.31
5.00	0.291	0.32
5.50	0.308	0.34
6.00	0.324	0.36
6.50	0.339	0.37
7.00	0.353	0.39
7.50	0.368	0.40
8.00	0.381	0.42
8.50	0.395	0.43
9.00	0.408	0.45
10.00	0.433	0.48
12.00	0.481	0.53
15.00	0.545	0.60
20.00	0.642	0.71
30.00	0.806	0.89
45.00	1.012	1.11
60.00	1.188	1.31

PVC 63 - ID 57mm- PN 10		
J (m/km)	Q (l/s)	v m/s)
0.50	0.317	0.12
0.60	0.353	0.14
0.70	0.385	0.15
0.80	0.416	0.16
0.90	0.446	0.17
1.00	0.474	0.19
1.10	0.500	0.20
1.20	0.526	0.21
1.30	0.551	0.22
1.40	0.575	0.23
1.50	0.598	0.23
1.60	0.620	0.24
1.70	0.642	0.25
1.80	0.664	0.26
1.90	0.684	0.27
2.00	0.705	0.28
2.25	0.754	0.30
2.50	0.801	0.31
2.75	0.845	0.33
3.00	0.888	0.35
3.25	0.929	0.36
3.50	0.969	0.38
3.75	1.008	0.40
4.00	1.046	0.41
4.25	1.082	0.42
4.50	1.118	0.44
4.75	1.153	0.45
5.00	1.187	0.47
5.50	1.252	0.49
6.00	1.315	0.52
6.50	1.376	0.54
7.00	1.435	0.56
7.50	1.492	0.58
8.00	1.547	0.61
8.50	1.601	0.63
9.00	1.653	0.65
10.00	1.754	0.69
12.00	1.942	0.76
15.00	2.200	0.86
20.00	2.583	1.01
30.00	3.236	1.27
45.00	4.049	1.59
60.00	4.744	1.86

PVC 63 - ID 53.6mm- PN 16		
J (m/km)	Q (l/s)	v (m/s)
0.50	0.268	0.12
0.60	0.298	0.13
0.70	0.326	0.14
0.80	0.352	0.16
0.90	0.377	0.17
1.00	0.400	0.18
1.10	0.423	0.19
1.20	0.445	0.20
1.30	0.466	0.21
1.40	0.486	0.22
1.50	0.506	0.22
1.60	0.525	0.23
1.70	0.543	0.24
1.80	0.561	0.25
1.90	0.579	0.26
2.00	0.596	0.26
2.25	0.638	0.28
2.50	0.677	0.30
2.75	0.715	0.32
3.00	0.752	0.33
3.25	0.787	0.35
3.50	0.820	0.36
3.75	0.853	0.38
4.00	0.885	0.39
4.25	0.916	0.41
4.50	0.946	0.42
4.75	0.976	0.43
5.00	1.005	0.45
5.50	1.060	0.47
6.00	1.114	0.49
6.50	1.165	0.52
7.00	1.215	0.54
7.50	1.263	0.56
8.00	1.310	0.58
8.50	1.356	0.60
9.00	1.400	0.62
10.00	1.486	0.66
12.00	1.646	0.73
15.00	1.865	0.83
20.00	2.190	0.97
30.00	2.744	1.22
45.00	3.435	1.52
60.00	4.025	1.78

PVC 90 - ID 81.4mm- PN 10		
J (m/km)	Q (l/s)	v (m/s)
0.50	0.841	0.16
0.60	0.933	0.18
0.70	1.019	0.20
0.80	1.100	0.21
0.90	1.177	0.23
1.00	1.250	0.24
1.10	1.319	0.25
1.20	1.386	0.27
1.30	1.451	0.28
1.40	1.513	0.29
1.50	1.574	0.30
1.60	1.632	0.31
1.70	1.689	0.32
1.80	1.745	0.34
1.90	1.799	0.35
2.00	1.852	0.36
2.25	1.980	0.38
2.50	2.101	0.40
2.75	2.217	0.43
3.00	2.329	0.45
3.25	2.436	0.47
3.50	2.540	0.49
3.75	2.640	0.51
4.00	2.738	0.53
4.25	2.833	0.54
4.50	2.925	0.56
4.75	3.015	0.58
5.00	3.103	0.60
5.50	3.273	0.63
6.00	3.436	0.66
6.50	3.594	0.69
7.00	3.746	0.72
7.50	3.893	0.75
8.00	4.035	0.78
8.50	4.174	0.80
9.00	4.309	0.83
10.00	4.570	0.88
12.00	5.057	0.97
15.00	5.723	1.10
20.00	6.710	1.29
30.00	8.389	1.61
45.00	10.478	2.01
60.00	12.260	2.36

PVC 90 - ID 76.6mm- PN 16		
J (m/km)	Q (l/s)	v (m/s)
0.50	0.712	0.15
0.60	0.791	0.17
0.70	0.864	0.19
0.80	0.933	0.20
0.90	0.998	0.22
1.00	1.060	0.23
1.10	1.119	0.24
1.20	1.176	0.26
1.30	1.230	0.27
1.40	1.283	0.28
1.50	1.335	0.29
1.60	1.385	0.30
1.70	1.433	0.31
1.80	1.480	0.32
1.90	1.526	0.33
2.00	1.571	0.34
2.25	1.680	0.36
2.50	1.783	0.39
2.75	1.882	0.41
3.00	1.976	0.43
3.25	2.068	0.45
3.50	2.156	0.47
3.75	2.241	0.49
4.00	2.324	0.50
4.25	2.405	0.52
4.50	2.483	0.54
4.75	2.560	0.56
5.00	2.635	0.57
5.50	2.779	0.60
6.00	2.918	0.63
6.50	3.052	0.66
7.00	3.181	0.69
7.50	3.306	0.72
8.00	3.428	0.74
8.50	3.546	0.77
9.00	3.661	0.79
10.00	3.882	0.84
12.00	4.297	0.93
15.00	4.864	1.06
20.00	5.704	1.24
30.00	7.133	1.55
45.00	8.912	1.93
60.00	10.429	2.26

PVC 110 - ID 101.6mm- PN 10		
J (m/km)	Q (l/s)	v (m/s)
0.50	1.536	0.19
0.60	1.705	0.21
0.70	1.861	0.23
0.80	2.008	0.25
0.90	2.146	0.26
1.00	2.278	0.28
1.10	2.405	0.30
1.20	2.526	0.31
1.30	2.643	0.33
1.40	2.756	0.34
1.50	2.865	0.35
1.60	2.971	0.37
1.70	3.075	0.38
1.80	3.175	0.39
1.90	3.273	0.40
2.00	3.369	0.42
2.25	3.600	0.44
2.50	3.819	0.47
2.75	4.029	0.50
3.00	4.231	0.52
3.25	4.425	0.55
3.50	4.612	0.57
3.75	4.793	0.59
4.00	4.970	0.61
4.25	5.141	0.63
4.50	5.307	0.65
4.75	5.470	0.67
5.00	5.629	0.69
5.50	5.936	0.73
6.00	6.230	0.77
6.50	6.514	0.80
7.00	6.788	0.84
7.50	7.053	0.87
8.00	7.310	0.90
8.50	7.560	0.93
9.00	7.803	0.96
10.00	8.272	1.02
12.00	9.150	1.13
15.00	10.349	1.28
20.00	12.124	1.50
30.00	15.142	1.87
45.00	18.891	2.33
60.00	22.088	2.72

PVC 110 - ID 96.8mm- PN 16		
J (m/km)	Q (l/s)	v (m/s)
0.50	1.347	0.18
0.60	1.495	0.20
0.70	1.632	0.22
0.80	1.761	0.24
0.90	1.883	0.26
1.00	1.999	0.27
1.10	2.110	0.29
1.20	2.216	0.30
1.30	2.319	0.32
1.40	2.418	0.33
1.50	2.514	0.34
1.60	2.608	0.35
1.70	2.698	0.37
1.80	2.787	0.38
1.90	2.873	0.39
2.00	2.957	0.40
2.25	3.160	0.43
2.50	3.353	0.46
2.75	3.537	0.48
3.00	3.714	0.50
3.25	3.885	0.53
3.50	4.049	0.55
3.75	4.209	0.57
4.00	4.364	0.59
4.25	4.514	0.61
4.50	4.661	0.63
4.75	4.804	0.65
5.00	4.943	0.67
5.50	5.213	0.71
6.00	5.472	0.74
6.50	5.722	0.78
7.00	5.962	0.81
7.50	6.196	0.84
8.00	6.422	0.87
8.50	6.642	0.90
9.00	6.856	0.93
10.00	7.268	0.99
12.00	8.040	1.09
15.00	9.095	1.24
20.00	10.656	1.45
30.00	13.312	1.81
45.00	16.612	2.26
60.00	19.426	2.64

PVC 160 - ID 147.6mm- PN 10		
J (m/km)	Q (l/s)	v (m/s)
0.50	4.222	0.25
0.60	4.680	0.27
0.70	5.104	0.30
0.80	5.503	0.32
0.90	5.879	0.34
1.00	6.238	0.36
1.10	6.580	0.38
1.20	6.909	0.40
1.30	7.226	0.42
1.40	7.532	0.44
1.50	7.828	0.46
1.60	8.116	0.47
1.70	8.395	0.49
1.80	8.668	0.51
1.90	8.933	0.52
2.00	9.193	0.54
2.25	9.816	0.57
2.50	10.409	0.61
2.75	10.976	0.64
3.00	11.520	0.67
3.25	12.044	0.70
3.50	12.550	0.73
3.75	13.039	0.76
4.00	13.514	0.79
4.25	13.976	0.82
4.50	14.425	0.84
4.75	14.863	0.87
5.00	15.291	0.89
5.50	16.118	0.94
6.00	16.911	0.99
6.50	17.675	1.03
7.00	18.412	1.08
7.50	19.125	1.12
8.00	19.817	1.16
8.50	20.490	1.20
9.00	21.144	1.24
10.00	22.404	1.31
12.00	24.761	1.45
15.00	27.979	1.64
20.00	32.738	1.91
30.00	40.818	2.39
45.00	50.839	2.97
60.00	59.371	3.47

PVC 160 - ID 141mm- PN 16		
J (m/km)	Q (l/s)	v (m/s)
0.50	3.732	0.24
0.60	4.136	0.26
0.70	4.512	0.29
0.80	4.865	0.31
0.90	5.198	0.33
1.00	5.515	0.35
1.10	5.818	0.37
1.20	6.110	0.39
1.30	6.390	0.41
1.40	6.661	0.43
1.50	6.923	0.44
1.60	7.178	0.46
1.70	7.426	0.48
1.80	7.667	0.49
1.90	7.902	0.51
2.00	8.131	0.52
2.25	8.684	0.56
2.50	9.209	0.59
2.75	9.711	0.62
3.00	10.193	0.65
3.25	10.656	0.68
3.50	11.104	0.71
3.75	11.538	0.74
4.00	11.959	0.77
4.25	12.367	0.79
4.50	12.765	0.82
4.75	13.154	0.84
5.00	13.532	0.87
5.50	14.265	0.91
6.00	14.968	0.96
6.50	15.644	1.00
7.00	16.298	1.04
7.50	16.930	1.08
8.00	17.543	1.12
8.50	18.138	1.16
9.00	18.718	1.20
10.00	19.835	1.27
12.00	21.924	1.40
15.00	24.775	1.59
20.00	28.994	1.86
30.00	36.156	2.32
45.00	45.043	2.88
60.00	52.609	3.37

PVC 200 - ID 184.6mm- PN 10		
J (m/km)	Q (l/s)	v (m/s)
0.50	7.714	0.29
0.60	8.545	0.32
0.70	9.316	0.35
0.80	10.04	0.38
0.90	10.723	0.4
1.00	11.373	0.42
1.10	11.995	0.45
1.20	12.591	0.47
1.30	13.166	0.49
1.40	13.721	0.51
1.50	14.258	0.53
1.60	14.779	0.55
1.70	15.286	0.57
1.80	15.779	0.59
1.90	16.26	0.61
2.00	16.73	0.63
2.25	17.86	0.67
2.50	18.934	0.71
2.75	19.96	0.75
3.00	20.944	0.78
3.25	21.892	0.82
3.50	22.807	0.85
3.75	23.692	0.89
4.00	24.551	0.92
4.25	25.386	0.95
4.50	26.198	0.98
4.75	26.99	1.01
5.00	27.763	1.04
5.50	29.258	1.09
6.00	30.691	1.15
6.50	32.071	1.2
7.00	33.402	1.25
7.50	34.691	1.3
8.00	35.94	1.34
8.50	37.154	1.39
9.00	38.335	1.43
10.00	40.609	1.52
12.00	44.862	1.68
15.00	50.665	1.89
20.00	59.242	2.21
30.00	73.791	2.76
45.00	91.821	3.43
60.00	107.159	4.00

PVC 200 - ID 176.2mm- PN 16		
J (m/km)	Q (l/s)	v (m/s)
0.50	6.805	0.28
0.60	7.539	0.31
0.70	8.221	0.34
0.80	8.86	0.36
0.90	9.463	0.39
1.00	10.038	0.41
1.10	10.587	0.43
1.20	11.114	0.46
1.30	11.621	0.48
1.40	12.112	0.5
1.50	12.586	0.52
1.60	13.047	0.54
1.70	13.495	0.55
1.80	13.931	0.57
1.90	14.356	0.59
2.00	14.771	0.61
2.25	15.769	0.65
2.50	16.719	0.69
2.75	17.625	0.72
3.00	18.496	0.76
3.25	19.333	0.79
3.50	20.142	0.83
3.75	20.925	0.86
4.00	21.684	0.89
4.25	22.422	0.92
4.50	23.14	0.95
4.75	23.84	0.98
5.00	24.524	1.01
5.50	25.846	1.06
6.00	27.113	1.11
6.50	28.333	1.16
7.00	29.51	1.21
7.50	30.65	1.26
8.00	31.754	1.3
8.50	32.828	1.35
9.00	33.872	1.39
10.00	35.88	1.47
12.00	39.65	1.63
15.00	44.78	1.84
20.00	52.37	2.15
30.00	65.24	2.68
45.00	81.20	3.33
60.00	94.77	3.89

B. FRICTION LOSS TABLE. GALVANIZED IRON

Approximate values for head loss in m/km of galvanized iron pipe, calculated using the Hazen-Williams formula for middle-aged pipe.

Important: head loss varies somewhat from one manufacturer to another. If a manufacturer provides you with reliable data, use this instead.

Flow	1/2"	1"	1 1/2"	2"	3"	4"	5"	6"
l/s	**15mm**	**25mm**	**40mm**	**50mm**	**80mm**	**100mm**	**125mm**	**150mm**
0.02	2.28							
0.05	12.46	1.22						
0.1	45.00	4.39						
0.15	95.34	9.31	0.94					
0.2	162.44	15.86	1.61					
0.25	245.56	23.97	2.43					
0.3	344.19	33.60	3.41	1.15				
0.35	457.92	44.71	4.53	1.53				
0.4	586.40	57.25	5.80	1.96				
0.45	729.33	71.20	7.22	2.43				
0.5	886.48	86.55	8.77	2.96				
0.6		121.31	12.30	4.15				
0.7		161.39	16.36	5.52				
0.8		206.67	20.95	7.07				
0.9		257.05	26.06	8.79				
1		312.43	31.67	10.68	0.92			
1.1		372.75	37.79	12.75	1.10			
1.2		437.93	44.40	14.98	1.29			
1.3		507.90	51.49	17.37	1.50			
1.4		582.62	59.06	19.92	1.72			
1.5		662.03	67.11	22.64	1.95			
1.6		746.08	75.63	25.51	2.20			
1.7			84.62	28.55	2.46			
1.8			94.07	31.73	2.74			
1.9			103.98	35.07	3.03	1.02		
2			57.13	38.57	3.33	1.12		
2.2			64.38	46.02	3.97	1.34		

Head loss values in m/km

Values calculated without velocity adjustments (valid for velocities of less than 3 m/s). Coefficient C-110 used for pipe of less than 3" diameter, and C-120 for 3" and over, for middle aged pipe with neutral water (Langelier Index value of ± 0.5).

l/s	15mm	25mm	40mm	50mm	80mm	100mm	125mm	150mm
Flow	**1/2"**	**1"**	**1 1/2"**	**2"**	**3"**	**4"**	**5"**	**6"**
2.4			72.03	54.06	4.66	1.57		
2.6			80.07	62.70	5.41	1.83		
2.8			88.50	71.92	6.21	2.09		
3			97.32	81.73	7.05	2.38		
3.2			116.11	92.10	7.95	2.68		
3.4			136.41	103.05	8.89	3.00	1.01	
3.6			158.21	114.55	9.88	3.33	1.12	
3.8			181.49	126.62	10.93	3.69	1.24	
4			206.22	139.24	12.01	4.05	1.37	
4.5			232.41	173.18	14.94	5.04	1.70	
5				210.49	18.16	6.13	2.07	
5.5				251.13	21.67	7.31	2.47	1.01
6				295.04	25.46	8.59	2.90	1.19
6.5				342.18	29.53	9.96	3.36	1.38
7					33.87	11.43	3.85	1.59
8					43.37	14.63	4.94	2.03
9					53.94	18.20	6.14	2.53
10					65.57	22.12	7.46	3.07
11					78.23	26.39	8.90	3.66
12					91.90	31.00	10.46	4.30
15					138.94	46.87	15.81	6.51
20					236.70	79.84	26.93	11.08
25					357.83	120.70	40.72	16.76
30						169.19	57.07	23.49
40						288.24	97.23	40.01
50						435.75	146.99	60.49

To calculate intermediate values, you can use the Hazen-Williams formula, taking into account the warnings and values detailed in the box:

$$h = \frac{10,7 L Q^{1,852}}{C^{1,852} D^{4,87}}$$

Where: h, head loss in meters; L, length in meters; C, friction coeffcient and D, diamater in meters and Q flow in m^3/s.

Version 1.0

CPSIA information can be obtained
at www.ICGtesting.com
Printed in the USA
LVHW050936130722
723396LV00009B/396